Handbook of Domestic Ventilation

Handbook of Domestic Ventilation

Rodger Edwards

ELSEVIER
BUTTERWORTH
HEINEMANN

AMSTERDAM • BOSTON • HEIDELBERG • LONDON • NEW YORK • OXFORD •
PARIS • SAN DIEGO • SAN FRANCISCO • SINGAPORE • SYDNEY • TOKYO

Elsevier Butterworth-Heinemann
Linacre House, Jordan Hill, Oxford OX2 8DP
30 Corporate Drive, Burlington, MA 01803

First published 2005

British Library Cataloguing in Publication Data
A catalogue record for this book is available from the British Library

Library of Congress Cataloguing in Publication Data
A catalogue record for this book is available from the Library of Congress

ISBN 0 7506 5097 4

For information on all Elsevier Butterworth-Heinemann publications visit our website at http://books.elsevier.com

Typeset by Charon Tec Pvt. Ltd, Chennai, India
www.charontec.com
Printed and bound in Great Britain

Contents

Foreword

There are some excellent textbooks about ventilation currently in print. With due respect to all other authors, those written by Hassam Awbi, David Etheridge and Mats Sandberg immediately spring to mind. Many aspects of ventilation are dealt with in the standard building services engineering textbooks. Although the bias tends to be towards mechanical ventilation. The one feature about all these sources is that there is little if any dedicated coverage of the ventilation of dwellings. This is most surprising, given that dwellings form such a large proportion of the UK's building stock. A common view is that ventilating houses is not complicated, and therefore there is no real point in getting too interested in the subject. This view is not consistent with the current state of the housing stock with respect to the incidence of condensation problems. We still have not got it right. Mainland Europe does not suffer from problems to the same extent as the UK. Climate differences are but part of the answer. This would suggest that the whole issue is rather more complicated than some people would have us believe.

It was a major surprise to me that nobody had bothered to write a textbook about domestic ventilation, aimed at a wide range of readers. I feel that such a book is long overdue.

Whilst there is currently no main textbook, there are many diverse sources of information about domestic ventilation. These range from Building Research Establishment reports and Digests through to research reports by a range of other organisations. There are elements of useful information within numerous other publications such as the CIBSE Guides. Given the implications for occupant health of inadequate ventilation, it comes as no surprise that some material can be located within the vast amounts of paper published within the environmental health sector. The majority of this information is very well written, and can be cited directly in a book of this nature with little by way of comment. The frustrating thing is that all this information has never been collated and cross-referenced in anything other format save that of the literature review of research theses – hardly what is required for general use. This book intends to remedy this problem. However, its scope goes beyond that of a mere literature search. I have been involved in several pieces of research related to domestic ventilation over the past 22 years. These range from the development of air movement measurement techniques to the monitoring of the performance of ventilation systems. This research will be cited within the book as appropriate. In several areas, for example with respect to the debate about the merits of passive stack ventilation versus mechanical ventilation, the findings will hopefully be of great interest to the reader.

Acknowledgements

A large number of researchers, legislators and others have been responsible for generating the significant amount of relevant literature collated and analysed in this book. It is perhaps a sign of the times that many of the people concerned, and for that matter some of the organisations, are no longer involved in domestic ventilation research. My apologies to anybody that I have unintentionally omitted from the list.

My thanks go to Prof. Peter Burberry, Chris Irwin, Ken Letherman, Stephen Cooper, David Pitt, Neil Rideout and all the other staff past and present at Building Product Design; Earle Perera, Roger Stephen and all other staff past and present at BRE who have made such notable contributions in the field; Andrew Gaze (ex TRADA); Ken Johnson (ex Pilkingtons); Martin Liddament (ex AIVC and now Veetech).

1

Introduction

Whilst the reasons for having a book about ventilation are fairly clear, perhaps this might not, at the first sight, seem to be the case for dwellings. From the perspective of building services engineering, houses are small and relatively of a small range of room configurations. The amount of air required for ventilation purposes is correspondingly small, and the measures required to ensure effective ventilation might be viewed as being relatively simple to implement.

This seeming simplicity does not reconcile with the statistics. A series of English House Condition Surveys reveal the ongoing issues with the performance of the housing stock. As the total stock level approaches the 20 million mark, condensation and mould growth problems are reported within about 15% of properties. This percentage has shown very little improvement over the past 20 years. This is despite a succession of amendments to the Building Regulations spanning almost three decades. Browsing through a box of books left behind at the University of Manchester Institute of Science and Technology (UMIST) by a retired colleague, the author came across a small book about condensation. Edited by Derek Croome and Alan Sherratt, two very familiar names to those in building services engineering circles, the book consisted of a series of chapters about various aspects of what were clearly felt, at that time (1972) be a significant problem. What were the main messages? Well, most of the issues that we would recognise today were discussed, albeit within the context of the construction methods and technologies of the day. One contributor, Phyllis Allen, mentioned about the changes in occupant behaviour and lifestyle that were contributing to condensation problems. Alec Loudon and others talked in some detail about ventilation requirements for condensation control, and also presented calculations that showed how much heat input would be required for a given property under selected conditions. Loudon was clearly aware of the fact that the possibility of over-ventilation is actually making condensation worse. Low indoor temperatures and inadequate heating were agreed to be the prime factors. One final thing that was quite clear is that with respect to energy conservation, little was understood by the significance of air leakage. To be fair, there was probably not much need.

Reading through the book, it becomes quite apparent that the major causes of condensation were all too well understood over 30 years ago, and in addition, so was the need for a balance between indoor temperatures and ventilation rate. The question might reasonably be, "why wasn't the available knowledge taken on board by the regulators of the day?"

In a very short time, after the publication of Croome and Sherratt's book, the "civilised world" spiralled off into its era of high oil prices and fuel shortages. Energy saving suddenly became fashionable. However, from the condensation perspective, many of the measures taken served to increase the incidences of condensation-related problems. To make it worse, the problems were actually extended into a part of the dwelling that thus far had not had any. A combination of changes served to produce a new difficulty, namely roof-space condensation. During the late 1970s, the UK construction industry discovered the virtues of timber-framed construction. Unfortunately, it did not understand the hygrothermal implications of using this method of construction within the damp environment prevailing in the UK. The consequence of this was an epidemic of interstitial condensation and structural damage. Since then, over the period of time, the UK legislators have been chasing the tail of the condensation monster, with only limited success at best. Setting aside issues of how appropriate regulatory changes have been formulated and applied, there are big problems regarding human factors with dwellings. Whatever is being designed into a dwelling, someone somewhere will conspire to render it ineffective or inoperable. Ventilation systems are particularly vulnerable in this respect. The author has seen many condensation problems caused by blocked off air inlets and outlets, and switched off or even deliberately damaged fans. In conclusion, one more dip into the closing remarks in Croome and Sherratt's book reveals the following priceless quotation:

> "... condensation seems rather like sex – if we cannot eliminate it the very least we can do is try to control it."

Perhaps elimination of condensation is really an unrealistic goal. Still this should not stop it from being set as an ideal objective.

This book is deliberately written so as to be of use to the widest possible readership. The practical implication of this strategy is that the inclusion of unnecessary mathematics is deliberately avoided. There are other books (and very good in one or more aspects) available, dealing with the mathematical aspects of ventilation, and there would be no point in reproducing such well-written theoretically based material in this book. However, in some chapters (e.g. Chapters 3 and 4), usage has been made of simple formulae. In such instances, every effort is made to present numerical examples.

The structure of the book is as follows. Chapter 2 describes the ventilation requirements within the dwellings, and links them to a range of issues relating to human health and comfort. Chapter 3 reviews a range of prediction procedures, not only for the calculation of ventilation requirements for a range of situations, but also for the prediction of air change rates within dwellings using predetermined environmental data. A brief mention is made about the calculation movements between interconnected cells within the dwellings. Chapter 4 examines the various experimental techniques that can be used for making *in situ* measurements of envelope airtightness, air change rates and other parameters. Chapter 5 examines the strategies that may be used for the ventilation of dwellings, whilst Chapter 6 is concerned with the specialised aspects of ventilation provision (such as the measures required in roof spaces), for flats and for the control of specific risks (such as radon and landfill gas ingress). In Chapter 7, the numerous (and rapidly increasing in number) regulatory aspects related to domestic ventilation are described, and the possible means of compliance are discussed. The information presented are not merely confined to Approved Document F, but also cover aspects of other approved documents relevant to domestic ventilation. Separate consideration of requirements for both, Scotland and Northern Ireland, is presented. As a conclusion for this book, Chapter 8 takes a semi-speculative look at the likely future trends in domestic ventilation.

2

Indoor Issues and Health Implications (and Ventilation Requirements)

2.1 Introduction

Within dwellings, the range of indoor environmental issues is almost as broad as that encountered in commercial and industrial premises. However, the significance of a given issue may vary from the commercial and industrial contexts.

The health implications of the indoor environment can be very far reaching. This is not unreasonable considering the amount of time that most people spend within their dwelling.

In this chapter, the key issues linked to indoor environment and those influenced by ventilation are discussed, and the main health implications are explored. At the end of this chapter, current thinking regarding recommended ventilation rates is summarised.

2.2 The need for ventilation

There are three main reasons to ventilate a building:

Firstly, ventilation is needed to provide adequate air for respiration purposes. Unless a building is made particularly airtight, the provision of air for this purpose is not likely to pose any problem. At present, such buildings do not exist, although some extremely high levels of airtightness are being claimed for some of the new modular type constructions (which is more detailed in Chapter 8) which might give rise to concerns if the purpose provided by means of ventilation is not functioning correctly.

Secondly, ventilation is required in order to provide cooling effect in the summer. Given the occupancy patterns typically encountered in dwellings, not much attention is paid to the cooling of dwellings in comparison to commercial buildings such as offices. Window opening, in offices, is generally deemed to be adequate, and within the UK at any rate there seems to be, but a limited scientific rationale, behind the provision made. Air conditioning or comfort cooling is provided only on the rarest of occasions, presumably on the grounds of cost, although its use is quite common in the US and particularly in apartment blocks.

Finally, the most important reason for the provision of ventilation within dwellings is for the removal and dilution of airborne contaminants. Of these, the most important one is water vapour, although there are others that are worthy of consideration.

2.3 Water vapour

Water vapour is present in ambient air. Its presence is part of nature itself, being a by-product of the metabolism of most forms of life. The amount of water vapour present in the external air is instrumental in determining whether rain and a range of weather phenomena take place. Readers will no doubt remember the "water cycle" diagram from their early days of science at school. It is not proposed to expound at length about the properties of water vapour within atmospheric air at this juncture. Suffice to say that in the usual range of environmental conditions the ability of air to carry water vapour is primarily determined by its temperature. The psychrometric chart shows that water vapour content can vary between dry (less than 0.005 kg/kg of dry air) in cold conditions (that is why freeze drying takes place) and quite moist (greater than 3.5 kg/kg of dry air) under typical summer conditions (which is why summer showers can be so torrential). For much more detailed information concerning water vapour, the psychrometric chart and its use, refer to Chapter 3.

Water vapour production as a consequence of human metabolism is a significant source within the indoor environment. Figure 2.1 shows the variation in moisture production rate with the level of activity. It can be seen that appreciable amounts of water vapour are generated during sleep, which is important when condensation risk within bedrooms is being considered. Other activities within the dwelling can also result in the production of large amounts of water vapour. Figure 2.2 shows typical rates of water vapour generation for a variety of activities. Some of the amounts of water vapour generated may seem to be surprising at the first

Daily moisture generation rates for households			
Number of persons in household	Daily moisture generation rates		
	Dry occupancy* (kg)	Moist occupancy† (kg)	Wet occupancy‡ (kg)
1	3.5	6	9
2	4	8	11
3	4	9	12
4	5	10	14
5	6	11	15
6	7	12	16

* Where there is proper use of ventilation, it includes those buildings unoccupied during the day; results in an internal vapour pressure up to 0.3 kPa in excess of the external vapour pressure.

† Where internal humidities are above normal; likely to have poor ventilation; possibly a family with children, water vapour excess in between 0.3 and 0.6 kPa.

‡ Ventilation hardly ever used; high moisture generation; probably a family with young children, water vapour pressure excess is greater than 0.6 kPa.

Figure 2.1 Variation in moisture production with activity level (BS5250)[1]

sight. It is estimated that a daily total water vapour production of 15 kg might well be expected in a house occupied by a large family.

Gas combustion is also a source of water vapour production. This source is the most important one when flueless appliances are considered. A particularly bad example is the portable Calor gas heater beloved of static caravan owners for many years. Such appliances sometimes find their way into houses, with disastrous consequences in terms of condensation. However, gas combustion carries a much more serious risk than water vapour, as will be seen from Section 2.6.

2.4 Mould growth

Moulds make up a large family of living organisms. It would be neither realistic nor useful to fully discuss their nature and diversity within the context of a book about domestic ventilation. The material presented will, therefore, be confined to that directly relevant to the moulds that grow under suitable (or do we mean *unsuitable*?) conditions within dwellings. Apologies to any microbiologists picking up this book.

The nature of moulds makes them the most adept to thriving on surfaces. The moulds take the form of a filament-like structure called the *mycelium*, which consists of a branched system of walled tubes called

Typical moisture generation rates for household activities	
Household activity	Moisture generation rate
People asleep active	 40 g/h per person 55 g/h per person
Cooking electricity gas	 2000 g/day 3000 g/day
Dishwashing	400 g/day
Bathing/washing	200 g/person per day
Washing clothes	500 g/day
Drying clothes indoor (e.g. using unvented tumble drier)	1500 g/person per day

Typical moisture generation rates from heating fuels	
Heating fuel	Moisture generation rate (g/kW h)
Natural gas*	150
Manufactured gas*	100
Paraffin	100
Coke*	30
Anthracite*	10
Electricity	0

* The majority of heating appliances using these fuels are ventilated to the outside air. Consequently the water vapour produced by combustion is not released directly into the dwelling.

Figure 2.2 Typical moisture generation rates for a range of activities (BS5250)[2]

hyphae. These hyphae grow over the host surface and enable the mould to absorb water by osmosis and to extract nutrients. The large surface-area-to-volume ratio of the mycelium means the uptake of moisture is very efficient; conversely, it also means that the structure may rapidly dry out under unsuitable conditions. The moulds can move water and nutrients around over a large area in a manner reminiscent of the transmission of electricity in the National Grid. In addition to water uptake by osmosis, moulds also produce water as a by-product of their metabolism.

Moulds reproduce by means of the production of spores. The spores are dispersed into the atmosphere either by air movement or by active discharge. Both internal and external air will carry spores of many different mould species. In the summer, the mould spore concentration in outdoor air is typically about 50,000 per cubic metre of air, but in the winter, this figure will decline to the order of hundreds per cubic metre.[3] In contrast, dwellings always have a source of mould spores within them. When spores are brought into the house, they will settle down on surfaces together with dust, only to be brought back into aerial suspension by such activities as cleaning. If moulds have become established within a dwelling, then they will constitute a direct source of spore input into the indoor air. In numerical terms, fungal spores are usually the single most common group of airborne spores. In many cases, they will outnumber airborne bacteria and pollen particles. The numbers of spores produced by a mould colony are enormous, as high as billions for a larger colony. Spores are resistant to desiccation, and are able to germinate over a wide range of environmental conditions. Once present, they are very difficult to displace.

Spores will germinate and grow if they land on a surface with suitable living conditions. Moulds use enzymes to convert material forming the surface upon which they are growing into food substances in a usable form. Many materials used within houses, such as timber, wallpaper and decorative fabric materials represent suitable sources of nutrition for moulds. While mould growth will often manifest itself as damage to decorative finishes, it should not be forgotten that the process of food conversion may cause actual structural damage, the most extreme case being fungal attack of timber structural elements, for example in roof spaces (refer to Chapter 6 for more information). Surface contaminants such as grease, can also provide a food supply.

Moulds and other fungi need moisture in order to germinate and grow. They also need nutrients, oxygen and favourable temperatures. In general, moulds will be able to live in a temperature range of 0–40°C. Below 0°C, many moulds may survive but will not grow or reproduce, whilst temperatures above 40°C will prevent them from growing and will eventually result in their death. The major implication of this is that the range of temperatures found within dwellings is ideal for virtually all known moulds. Whilst lower temperatures in the range slow down the growth of some moulds, in practice they will still be able to grow on inside walls at the temperatures likely to be found in houses affected by condensation during the British winter. Within a building, damp or condensation can provide a suitable source of moisture on a prolonged basis. With rising damp, it is very often the case that the dissolved salts carried up by the water may serve as an inhibitor to the success of mould propagation. This is an important factor that should be borne in mind when determining whether a particular problem in a dwelling is due to condensation or damp.

The key factor in determining the growth rate of moulds and other fungi is the availability of moisture at the surface in question. This is represented by the term *water activity*, and is to all intents and purposes the relative humidity at the surface divided by 100 (refer to Chapter 3 for more information on the concept of relative humidity). It used to be assumed that the deposition of liquid water was necessary before mould growth could take place. However, it is now understood that the relative humidity of air within a space has a direct influence upon the level of water activity in surfaces. Individual species of mould have differing requirements in terms of minimum water activity levels. An activity level in excess of 0.8 indicates that most moulds will be able to grow and propagate. This is widely accepted as being synonymous with a relative humidity in excess of 70% within a given room or dwelling; indeed, this is the expression of the threshold as used in BS5250.

Different types of mould do better at particular levels of moisture activity. For example, *Penicillium* species can prosper in relatively dry conditions. The success of a mould species on a surface under such conditions can promote colonisation and succession by other mould species needing wetter conditions.

Moulds and other fungi cannot photosynthesise (i.e. they cannot produce nutrients from carbon dioxide and water in the presence of sunlight) in the manner of plants and algae, and therefore have to obtain their nutrients saprophytically; in other words, as parasites. They are able to extract food from a wide range of carbon compounds. As long as the enzymes can break down a given substance, food can be extracted. Nitrogen is another important part of the food requirements of fungi. This may be taken from the organic material within the substrate, but equally could be derived from inorganic sources, such as ammonium or nitrate salts. The quantities of nutrients needed for moulds to grow and prosper are very small, and can in fact easily be derived from surface deposits of dust and grease, even in dwellings which are kept very clean. This also explains why moulds can grow on inorganic surfaces such as plaster.

The occupied space of a dwelling provides an ideal home for moulds; however, it should be borne in mind that moulds will not exclusively live within the occupied space (refer to Section 6.1 concerning roof-space condensation). As has been noted earlier, whilst the growth rates of mould are influenced by temperature, the vast majority of mould species will thrive in the range of temperatures found within the typical dwelling. The presence of liquid water caused by condensation is usually intermittent, varying with the rates of occupancy, moisture production and ventilation rate. However, such a regime is adequate for the needs of most types of moulds. The most favourable time of year for mould growth within dwellings is during the winter months. This is because rates of moisture production are higher, ventilation rates are lower because windows are opened less often, and the internal surfaces are cooler.

Hunter and Sanders[4] have carried out an extensive review of the species of mould found in houses that were affected by condensation. They found that of the many thousands of species of fungi known to exist, only a very small proportion of these, perhaps less than 100, are found within housing.

The amounts and levels of mould spores within dwellings vary considerably. A common representation of the number of microbes present in a given sample, as used by microbiologists, is the actual number of pieces of matter containing microbes and are which in turn capable of breeding and spreading. The actual term used is the colony forming unit, often denoted as the cfu. When examining a Petri dish for evidence of microbial growth, the microbiologist is often seeking to determine the number of cfu present. For example, in a case where contamination of food was suspected, the inference would be the higher the number of cfu, the higher the level of contamination. When airborne levels of microbes need to be quantified, air samples of a predetermined volume are passed through a collecting medium (possibly a sterile filter pad). Microbial cultures are then incubated in a medium supportive of the microbes of interest. The numbers of cfu are then counted. Airborne levels of microbial contamination are expressed as the number of cfu per cubic metre (cfu/m^3) of air. The words "level" and "concentration" are used interchangeably within the literature, and this practice is followed within this book.

The measured levels of spores reported in the literature vary considerably. Some works[5] show maximum levels of about 6000 cfu/m^3, whilst other studies[6] indicate maxima of between 13,000 and 20,000 cfu/m^3. The highest level reported in the literature is 450,000. In comparison with some measurements of spore levels reported in commercial and industrial process buildings, concentrations within dwellings are relatively low. For example, concentrations of 10^6 cfu/m^3 have been measured in a museum basement in which old books were being stored in damp conditions,[7] whilst spore concentrations in excess of 10^8 cfu/m^3 have been measured in industrial buildings where mouldy materials were being handled.[8]

As might be expected, some differences are observed when spore levels within houses with surface mould growth are compared with those that are not. Spore levels between 6 and 2200 cfu/m^3 have been reported in houses which are not visibly affected by moulds, compared with levels between 10,000 and 15,000 in houses that exhibited surface mould growth. Generally speaking, during the spring to autumn period, indoor spore levels vary with the outdoor spore concentration, but the former levels are lower. However, in the winter, indoor spore levels are usually higher than outside. This is a reflection of the favourable conditions for mould propagation and growth within dwellings during the winter period.

Movement of furniture or cleaning activities may cause large but temporary changes in spore concentrations. Increases of the order of 400-fold have been observed, depending on the proximity to the surface or location that is being cleaned.

In cases where there is no mould growth within a dwelling, the spore content of the indoor air is essentially dependent on that of the outside air. Activities such as cleaning may cause short transient peaks in spore concentrations, but typically, concentrations will be between 10% and 50% of the outdoor concentration. If extensive mould growth takes place within a dwelling, then elevated concentrations of spores will result. It is the spores rather than the mould itself that constitutes the hazard to human health. The risk is due to the production of allergic reactions on inhalation of the spores. There are two distinct classes of allergic reactions. These classes are not exclusively confined to the description of the effects of spore inhalation, but are widely used to describe the medical effects of a wide range of allergens. With an atopic reaction, sometimes referred to as a *Type I allergy*, subjects have a natural predisposition to become sensitised to spores at their normal atmospheric concentration. Sensitisation to the spores increases through childhood to adult levels. The tendency to suffer from sensitisation differs from person to person. The symptoms may include itching, wheezing, rhinitis, conjunctivitis and sneezing. The symptoms are unpleasant for the sufferer. However, if the allergen source is removed, the symptoms quickly disappear, and no permanent damage is caused to the sufferer. To cause an atopic reaction in an exposed subject, fungal spores must be greater than 3 μm in their narrowest dimension.

Whereas about half the adult population exhibits an allergic reaction caused by one or more of the three most common allergens (grass pollen, cat fur and dust mites – of which more later), allergy caused by mould spores seems to be much less common. Even in a group of high-risk subjects, such as asthmatics, the body of medical evidence suggests that generally not more than 20% will be allergic to spores. Whilst a range of moulds with allergenic properties have been identified within dwellings, no positive correlation has been established between the presence of specific moulds in the indoor environment and incidence of allergic reaction. The effect of mould spores is masked by a range of secondary factors, such as level of occupation and smoking. Therefore, it is difficult to assess the importance of mould spores with respect to their independent influence on the occurrence of Type I allergy. Raw and Hamilton[5] assert that among subject samples consisting of children, the best estimate is that wheezing is twice as likely to occur in dwellings that are reported as being severely affected by mould. Within adult sample populations, they consider that there is even less evidence *either* for or against a link between mould growth and asthma.

In contrast to atopic subjects, non-atopic (otherwise known as *Type III allergy*) subjects usually become sensitised only after either a prolonged period of exposure to spores or after exposure to spores at a high concentration. Medical evidence suggests that exposure to concentrations of spores in excess of 10^6 colonies per cubic metre of air will be sufficient to cause a problem. Once sensitised to the spores, a subject will react more slowly upon exposure to allergens than would an atopic subject. However, the symptoms, which may include coughing, breathlessness and fever, will usually last rather longer after the removal of the allergen source. This in itself is a difficulty. The most serious effect on non-atopic subjects is that the allergic reaction may cause a condition called *extrinsic allergic alveolitis*, which may result in lasting damage to the lungs of sufferers.

This is a potentially serious consequence of Type III exposure, and sets it above Type I exposure as a danger.

Potential susceptibility to Type III allergy seems to be universal amongst the population at large, which again is very different from the case with Type I allergy. However, Type I is the most common problem of the two within dwellings. Indeed, Type III allergy is predominantly an industrial disease within the UK. Probably, the best-known example is the so-called "farmers lung" condition, which is caused by an exposure to the spores present in mouldy hay.

As is the case with inanimate particulate materials, the size of spores shows a range of typically between 1 and 100 micrometres (μm, which is more commonly referred to as micron). The actual health problems that may arise which are linked to the size of the spores in question. Smaller spores have the same tendency to travel further into the respiratory system as do small inanimate particles. Spores of less than 4 microns size can travel into the lungs as far as the alveoli are concerned (blood–gas exchange sites) and may cause a condition called *alveolitis*. In contrast, spores between 4 and 10 microns in size will become trapped in the bronchi and bronchioles, and will act as a trigger for asthma. Spores of greater than 10 microns in size will usually be held within the nasal passages. These will produce rhinitis, although if inhaled through the mouth they will also reach the bronchioles.

In addition to the possibility of allergic effects caused by breathing in spores, it is possible that they may also be a source of infection if inhaled. However, instances of such infections are thought to be more of a problem amongst groups of susceptible people. For example, *Aspergillus* infections of the airways can be a serious problem for asthma sufferers, if the spores of the fungus are inhaled. Given the types of fungi present in dwellings within the UK, it can be reasonably said that there is no serious risk of infection to the vast majority of the population.

Effects due to mould and spores may not be merely allergic or infective in nature. Some moulds produce very toxic compounds which are known as *mycotoxins*. These may, for example, interfere with the immune system and leave sufferers open to secondary infections. For example, it has been shown that there is a correlation between incidences of stomach upsets and diarrhoea in children and mould growth in housing.[9] The exposure of skin to mycotoxins (toxic compounds produced by moulds and other fungi) may cause skin irritation and allergic reactions, although the latter will probably be of less importance than the Type I allergic reaction caused by the spores themselves. There is also some limited evidence that some mycotoxins (e.g. the aflotoxins) may have carcinogenic properties in addition to toxic properties. Those readers who feed the wild birds in their gardens will have noted the "aflotoxin-free" labels on the more expensive bags of peanuts at the garden centre.

In addition to the mycotoxins, a range of volatile organic compounds (VOCs) (refer to Section 2.8 for more about VOCs) are produced by spores as metabolic by-products. These compounds are for the most part short-chain alcohols and aldehydes. Exposure to these compounds could in principle cause reactions, such as headaches, irritation of the eyes, nose and throat, and fatigue. However, the actual health effects reported vary considerably. The severity of symptoms among occupants of affected dwellings ranges from no effect through to quite severe illness. This demonstrates that whereas certain VOCs can be identified as being associated with the presence of mould and spores, the actual effects upon health cannot be quantified with any certainty. It is suggested by the author that the rational approach would be not to consider these particular VOCs in isolation, but rather as one component of the overall VOC load within the dwelling in question.

Whilst it is only natural to look for direct evidence of physiological effects of mould growth upon the occupants of dwellings affected by mould growth, not all of the effects on human health will be of this type. The possibility of adverse psychological effects should not be forgotten. In fact, dampness and mould growth are two of the most well-established causes of mental distress due to housing conditions. Depression may be caused amongst subjects, and particularly among women. Occupants will feel frustrated, if they are unable to either take or cause actions to be taken which will end the problem. The smell of dampness and mould is all pervasive in badly affected properties, and the problem may be so severe as to discourage the occupants of affected properties from inviting visitors to their houses. Particularly severe problems may lead to mould damage to fixtures, fittings and personal belongings such as clothing. The reappearance of mould growth after re-decoration without remedial actions against the cause of the problem may be a cause of significant depression.

In summary, mould growth due to condensation (and to a lesser extent damp) is a very serious danger to both the physical and mental well-being of the occupants of affected dwellings, and therefore it is very important that the mould growth is prevented. It should also be stressed that remedial measures designed to remove mould such as fungicides (such as antifungicidal washes and paints) will only cause a temporary solution, and indeed may in themselves have undesirable health effects. In the absence of fundamental changes to the internal environment as a result of better ventilation, improved heating or decreased water vapour input, mould growth will merely return when the effectiveness of the fungicide has worn off.

2.5 House dust mites

Health effects due to mites and similar creatures are not merely an issue within dwellings. There are several workplace illnesses that are caused by such creatures. For example, agricultural workers and those workers in granaries who are in contact with grain may be affected by asthma as a consequence of the presence of the grain weevil, whilst other mites carried on the grain may cause dermatitis. Such problems can be controlled by good hygiene in the workplace, coupled with the use of suitable protective clothing and equipment.

Unfortunately, the control of house dust mites is not such an easy task.

The most common species of house dust mite within the UK is *Dematophagoides pteronyssinus*. The literal translation of this name means "skin-eating feather mite". Mites are in fact arachnids, and so are closely related to spiders. Unlike their cousins, however, they are typically less than one-third of a millimetre in body length. Therefore, at least it can be said that their capacity to directly terrify is non-existent! The house dust mite causes problems of a more serious nature.

The house dust mite has long been established as a significant source of allergens. Voorhorst[10] demonstrated that *D. pteronyssinus* derives its nutrition from human skin scales. These scales are steadily shed from the skins of human beings, and are, therefore, a major constituent of house dust. It is not the house dust mite itself that poses the risk to human health, but rather a substance found in its droppings. The Group I allergen widely referred to as *Derp I* has been shown by Taylor[11] to act as a trigger of symptoms within those suffering from asthma. Figure 2.3 gives recommended maximum exposure concentrations for exposure to Derp I and house dust mites as recommended by the World Health Organisation (WHO). The lower concentrations must not be exceeded in order that sensitisation of subjects and the development of asthma itself are avoided. If the upper concentrations given are exceeded, then it would be expected that an acute asthma attack would be triggered in most patients who were already allergically sensitised to Derp I.

The indoor environment has a great influence on the prospects of success for *D. pteronyssinus*. The house dust mite prospers best in a warm moist place in which house dust can build up. Favoured locations for mite colonies are, therefore, in carpets, soft furnishing, bedding and mattresses. The optimum ambient temperature for the mite is about 25°C; the optimum relative humidity similar to those associated with the occurrence of mould growth and spore propagation varies typically between 74% and 80%. At relative humidities less than 48% at 20°C, corresponding to a water vapour content of 7 kg/kg of dry air, mites do not seem to multiply. They also do not seem to do very well in houses at high altitude. Presumably this must have some linkage to lower water vapour contents.

	Sensitisation	Acute attack
μg Derp I/g of dust	2	10
mites/g of dust	100	500

Figure 2.3 House dust mites – WHO concentrations

Water vapour content seems to be the favoured control marker for house dust mites, and studies of the effectiveness of ventilation systems have focused on the control of water vapour contents to a value below this level. In view of the encouragement provided by the climate of the UK to the growth of mould within dwellings, it should come as no surprise that mites seem to find conditions to their liking as well. Any dust mite problems will be made worse by inadequate heating, damp and the other deficiencies associated with poor housing.

It is difficult to precisely quantify the effect of the house dust mite on the incidence of asthma. There are a number of outdoor air pollutants, and in particular those associated with increases in levels of road traffic, which are known to contribute to the risk of asthma. However, what is quite clear is that there are now about half a million asthma sufferers within the UK, and, more worryingly, it is estimated that about one in seven children of primary school age are now sufferers.[12] What can be said with some certainty is that diagnosed cases of asthma, especially amongst children, are on the increase. There are so many environmental and social factors influencing the risk of asthma associated with a given subject, among them allergies (reported cases of which also seem to be on the increase), diet, indoor and external environment (including environmental tobacco smoke). Therefore it would probably be unfair to blame the humble house dust mite for all the problems being encountered. Equally, it would be most unwise to ignore their significance.

In practical terms, there are some measures that may be adopted in order to combat the growth of dust mites and reduce the risk of exposure to Derp I. It has been reported by Owens[13] that the use of polyurethane-coated textiles on bedding and mattresses has been shown to reduce Derp I concentrations within the dwellings by about 99%. Dispensing with carpet and using smooth floor finishes will remove a favoured habitat of the dust mite. Psychrometric conditions within actual carpeting has not as yet been the subject of a detailed scientific investigation; but given the ability of fibrous materials to absorb and desorb water vapour, it is highly likely that suitable conditions for the success of house dust mites may exist even when the actual water vapour content in the room above is well below the 7 kg/kg dry air control value. Washing the bedding at a water temperature in excess of 58°C will both kill mites and remove existing build-ups of allergens. There are chemical agents available, which will kill mites. These are referred to as *acaricides*. Such agents will, of course, have no effect on the existing build-ups of Derp I, and their use may in itself be associated with health risks. The use of such agents may, as in the case of pesticides and other similar chemical agents, result in the production of resistant strains of mite in the long term, and so their use might prove to be counterproductive. There is a spray-on product available which neutralises allergens; however, its effect is likely to be short term unless applied regularly.[14] Whilst it might be thought that regular vacuum cleaning might remove dust mites and allergens, there is a danger that the airborne allergen concentration might be increased unless the vacuum cleaner used is fitted with a high-efficiency particle arrestance (HEPA) filter. This type of filtration equipment is of the type more commonly associated with clean rooms and pharmaceutical production facilities. Currently, there seems to be a lot of mass media advertisements centred around the virtues of HEPA filtration as a method of controlling dust mites. The HEPA filter may indeed have the capability of removing dust mites and the associated allergens from the environment; but there is a world of difference between being able to demonstrate large reductions in house dust mite numbers in pieces of carpet cleaned under controlled conditions and the reality of regularly cleaning a whole dwelling in the thorough manner required to have any realistic prospect of controlling the house dust mite population. In any case, not all potential homes for the mites lend themselves to vacuum cleaning. Within the context of their more established uses, HEPA filters are probably not subjected to the range of particle sizes that would be expected within the domestic situation. They are really intended to operate with suitable pre-filters for the removal of large particles and consequent protection of the HEPA filter medium.

In addition, given the critical consequences of product contamination (non-functioning expensive semiconductors and still worse, dangerous pharmaceutical products), HEPA filters, which are used as components of heating, ventilation and air conditioning (HVAC) systems servicing process plants, are operating within a regime of regular condition monitoring and filter medium replacement by trained staff. In the opinion of the author, the prospects of householders regularly replacing HEPA filter media in vacuum cleaners (let alone competently) are not very encouraging.

At present, there seems to be a growing interest in the science and technology of low-allergen housing. The ideas being expounded bring together such matters as appropriate means of ventilation, effective cleaning strategies and safe choices of materials for surfaces and soft furnishings. This area of development offers the potential of help for those who are particularly vulnerable to allergens, but is not cost effective for the housing sector at large.

The control of indoor air moisture content seems to provide the best prospect of successful dust mite control. As has been previously mentioned, 48% relative humidity at 20°C corresponds to a mixing ratio of 7 g/kg dry air. Scandanavian studies[15,16] indicate that ventilation is the preferred way of controlling moisture contents; furthermore mechanical supply and extract ventilation are associated with both lower mixing ratios and allergen concentrations. The winter climate in Scandanavia is both colder and drier than in the UK. Typically the mean mixing ratio in Denmark is about 1 g/kg lower than for the UK. The work of McIntyre[17] demonstrated that the use of mechanical ventilation resulted in lower mixing ratios than in naturally ventilated houses, and also a significant increase in the number of properties with a mean mixing ratio of less than 7 g/kg, 75% compared to 15% in the naturally ventilated group. However, the study by Fletcher *et al.*[18] of hygrothermal conditions in a group of nine mechanically ventilated houses shows mean mixing ratios of less than 7 g/kg were not maintained. In addition to this, house dust mite populations were not reduced. This latter observation was entirely consistent with previous finding about optimum conditions for house dust mite propagation. Therefore, it cannot be shown on the basis of the existing evidence that mechanical ventilation will provide a solution to house dust mite problems in all cases in the damp climatic conditions of the UK. Finally, the use of stand-alone dehumidifier units has been shown not to give any control over dust mite populations.[19]

The use of mechanical ventilation may possibly be a means of reducing the risk to health from house dust mites. However, there are extra capital costs incurred in the installation of mechanical ventilation systems, and energy costs will be increased. (The issue of energy implications of different ventilation strategies is discussed in more detail in Chapter 7.) As a result, there is little chance of mechanical ventilation being widely adopted as a control measure for house dust mites, given the inconclusive nature of the existing scientific data. This situation may change as dwellings become more airtight.

2.6 Carbon monoxide

A very significant source of pollutants within dwellings is the combustion of natural gas in a variety of appliances. In the days of town gas, the gas itself was contaminated with carbon monoxide as a by-product of its manufacture from coal. Natural gas is not contaminated in the same way. Indeed, the characteristic "smell of gas" is due to an additive rather than an intrinsic feature of the gas itself. Natural gas is typically 98.90% methane, 0.016% ethane, 0.02% propane, 0.87% nitrogen, and 0.02% of each of carbon dioxide and oxygen. Occasionally, there may be a supply of gas which has a trace contamination of hydrogen sulphide.

Under conditions of adequate oxygen supply, natural gas burns to form carbon dioxide, carbon monoxide and water vapour.

Of course, this is an exothermic reaction, as the reason for burning gas is to provide a source of heat. This is not the only reaction taking place (for more details about nitrogen oxides, refer to Section 2.8).

There is an increasing body of evidence which points towards the dangers of a gradual accumulation of carbon monoxide in the bloodstream. This occurs because the carbon monoxide molecule binds itself to haemoglobin to form carboxyhaemoglobin (often denoted in medical texts as COHb). Accumulations of carbon monoxide in the blood of smokers have been noted in studies of the health effects of smoking. Once within the bloodstream the disappearance of COHb takes place on a half-life basis: that is, it behaves in a similar manner to a radioactive isotope. In this case, the half-life is about 4 hours. If exposure to carbon monoxide is taking place during the day, as would probably be the case; this means that in practice the maximum blood concentration will be observed in the evening just before bedtime. During a night time period of 8 hours, the maximum level would decrease to about one-quarter of the maximum concentration. If no further exposure to the source were encountered, it would take over a day for the concentration to drop to 1% of the maximum. It is clear that regular exposure to even low levels of carbon monoxide can result in significant accumulations.

The formation of COHb within the bloodstream means that the supply of oxygen to the body is being reduced, because the blood cannot carry as much oxygen. The parts of the body needing the most oxygen to function correctly, most noticeably the brain and muscles, are the first to become affected. Symptoms may seem similar to influenza, and may include tiredness, dizziness, nausea, vomiting and diminution of mental faculties.[20] Prolonged exposure below a concentration which is actually fatal may manifest itself as a lingering influenza-like illness that shows no signs of improvement, and should be the foremost in the minds of general practitioners (GPs) if confronted with such a patient or indeed family of patients.

Steady accumulations of carbon monoxide in the medium term can lead to death. Figure 2.4 shows the relationship between percentage of COHb level and likely health effects arising. Even if the blood concentration of carbon monoxide remains below the level which is linked with death, there is evidence that strongly points towards elevated carbon monoxide concentration levels being associated with increased risk of early onset of angina and strokes.[22] New evidence suggests that there a number of possible other side effects. For example, it is now suggested that the temperamental behaviour of chefs is, in fact, attributable to prolonged exposure to carbon monoxide within the kitchen rather than any intrinsic character defect amongst those in the trade. As the basis of evidence increases, it is very likely that there will be a major change in attitudes towards ventilation provision for the control of carbon monoxide. In particular, the evidence about the effects of long-term low-level doses means that conventional concepts of merely restricting concentrations within spaces may no longer be tenable.

Balanced flue-type appliances such as modern boilers do not come with any need for ventilation air for combustion, of course, providing that they are being well maintained and the flue does not become disconnected, damaged or leaky. This is not the case for open flued appliances, such as the older types of instantaneous water heaters, gas cookers, gas fires and living flame-type fires. For these, the provision of an adequate supply of ventilation air is not merely an option, but rather it is essential. Failure to do so will result in the emission of potentially lethal amounts of carbon monoxide into the space. Approximately 1000 deaths per year are thought to be caused by faulty gas appliances and/or inadequate supply of ventilation air (refer to Chapter 7 for more details of regulatory requirements). The

% COHb in blood	Effects Associated with the COHb Level
80	Death[a]
60	Loss of consciousness; death if exposure continues[a]
40	Confusion; collapse on exercise[a]
30	Headache; fatigue; impaired judgement[a]
7–20	Statistically significant decreased maximal oxygen consumption during strenuous exercise in healthy young men[b]
5–17	Statistically significant diminution of visual perception, manual dexterity, ability to leam, or performance in complex sensorimotor tasks (such as driving)[b]
5–5.5	Statistically significant decreased maximal oxygen consumption and exercise time during strenuous exercise in young healthy men[b]
Below 5	No statistically significant vigilance decrements after exposure to CO[b]
2.9–4.5	Statistically significant decreased exercise capacity (i.e. shortened duration of exercise before onset of pain) in patients with angina pectoris and increased duration of angina attacks[b]
2.3–4.3	Statistically significant decreased (about 3–7%) work time to exhaustion in exercising healthy men[b]

[a] US EPA (1979); [b] US EPA (1985).

Figure 2.4 Percentage of COHb levels and related health effects (taken from reference[21])

majority of deaths occur in private rented sectors, where there has historically been reluctance on the part of some landlords to carry out any form of repair of maintenance. In 2000, in response to this serious health and safety issue, the government introduced a statutory scheme of gas appliance licensing and maintenance. There would be clear advantages in extending this scheme to householders, but this would be expensive to administer and indeed, given the current acute lack of registered gas installers, highly problematic to ensure that all tests and maintenance are carried out.

2.7 Carbon dioxide

Carbon dioxide is another by-product of gas combustion. It is also a by-product of respiration. Carbon dioxide occurs naturally in clean outdoor air at a concentration of about 0.03%. There are natural sources of carbon dioxide in addition to human respiration. For example, if acidic groundwater flows through limestone substrata, then carbon dioxide will be emitted, and it will eventually find its way to the surface. Carbon dioxide is also one of the constituents of landfill gas (refer to Section 2.10).

Carbon dioxide is a toxic gas. Exposure to concentrations above 6% will quickly prove to be fatal. Lower concentrations are associated with increased respiration rates, which might prove undesirable for those with breathing difficulties. Concentrations below the toxic limit may result in impaired judgement. The Health and Safety Executive[23] has set two maximum levels for exposure to carbon dioxide in commercial buildings and factories. The short-term limit, measured over a 20-min period, is 1.5%, whilst the occupational exposure limit, taken as the mean over an 8-h period, is 0.5%. Such concentrations are rarely experienced in dwellings and commercial buildings. It is generally accepted that the carbon dioxide level within an office building should not exceed 1000 parts per million (ppm) (0.1%), and it would be reasonable to take this value for dwellings. In practice, it is most unlikely that the concentration within the house will rise above this level if ventilation provision is adequate. In theory, high carbon dioxide concentrations could result from the combustion of natural gas; but in practice, the presence of carbon monoxide is a far more serious risk to deal with (for more details on air supply for gas appliances, refer to Chapter 7).

2.8 Nitrogen oxides

Even if an adequate amount of combustion air is supplied, this does not mean that gas combustion is free from the risk of pollutant emissions. The burning of gas results in the production of nitrous oxides. As was alluded to in Section 2.6, there is a secondary chemical reaction taking place, in which nitrogen (N_2) and oxygen (O_2) are combining to give a mixture of products which primarily consists of nitrogen dioxide (NO_2) and nitrogen trioxide (NO_3). This mixture is usually referred to as NO_x for convenience.

In the presence of water, NO_2 reacts to form nitric acid. This means that inhalation of NO_2 would result in the formation of nitric acid within the respiratory system itself. The extent of the chemical reaction is independent of the nitrogen content of the natural gas itself; the excess of nitrogen in ambient air overwhelming the former. Since the reaction is endothermic, the extent to which it takes place is a function of the actual combustion conditions. In practice, this relates to the design of the burner device at which gas combustion takes place.

On the basis of the available literature, the UK seems to lag behind some other countries with respect to the understanding of the significance of gas burner design. For instance, in Australia, rate of NO_x production has been measured for a number of existing gas burners.[24] Values as high as 60 nanograms per joule (ng/J) of energy from gas combustion have been recorded. The Australians have introduced a regulation which limits the amount of NO_x emission from new gas burners to 10 ng/J. Furthermore, a medium target of not greater than 5 ng/J has been set. The Australians regard this as being readily achievable: indeed, even 2 ng/J might be possible. It is clear that the Australian approach centres around controlling NO_x concentrations rather than devising ventilation strategies for control purposes.

Even with flues working correctly, there is evidence which points to spillage of other combustion pollutants into the space containing the appliance.[25,26] Spillage of emissions from cookers is seen as a particular issue. The provision of local extract ventilation for cookers is sometimes made, but is not at present mandatory. Oxides of nitrogen are known to exacerbate problems with asthma.

2.9 Other combustion by-products

In addition to nitrous oxides, there are a wide range of other emission products arising from cooking which are not associated with the use of gas. There is recent evidence which suggests that the cooking of food can result in the emission of substances hazardous to health. In extreme cases, these substances may be *carcinogenic* (cause cancer) or even *mutagenic* (cause genetic mutation). At present, these findings have not had any effect on legislation, but it is fairly certain that they must eventually have an impact. Document DW28, issued by the (HEVAC) Association,[27] makes reference to the potentially serious risks to health posed by the fumes produced by cooking. In view of the seriousness of the health consequences of these emissions, it is perhaps surprising that the more worrying aspects of this document have not been seized on by the media in the same way that other contemporary health scares have been.

2.10 Volatile organic compounds

The term *volatile organic compounds* (referred to as VOCs) is usually used to cover a wide range of chemical compounds with boiling points of about 250°C or less. Some other organic compounds with boiling points up to 400°C may also be present. The WHO defines a VOC as having a melting point below room temperature and a boiling point of between 50°C and 260°C.[28] In practice, this means that there may be a cocktail of several hundred VOCs in the air within a given building. In general, both indoor concentrations of VOCs and durations of exposure to them are higher indoor rather than outdoors.

Indoor VOCs can originate from a diverse range of sources. The most common of these are: man-made building materials and material treatments (such as damp proofing); decorations and furniture; equipment within buildings (e.g. photocopiers); emissions from outdoor sources (in particular, those from motor vehicles); metabolic by-products (not only from humans but also from plants and animals, including fungi); from the combustion of fuel and finally tobacco smoke. In some cases, VOCs may permeate into buildings from landfill sites.

When assessing the likely levels of VOCs within a building, it should be borne in mind that emission rates will not be constant during its lifespan. For example, emissions from building materials are usually at a peak during the period immediately after construction, and they will tail off to a very low emission rate in the fullness of time. The level of ventilation will have an influence on the rate at which emissions proceed. Low ventilation rates will tend to keep emission rates high for a longer period of time.

It is not sufficient merely to consider the rate at which VOCs are emitted. VOCs can be absorbed and desorbed from wallpaper and other soft furnishings in the same way as water vapour. The rates of absorption and desorption depend on the nature of the surface finishings, the VOC in question, together with internal environmental parameters such as temperature, relative humidity and ventilation rate.

Some VOCs may be present in sufficiently high concentrations as to cause bad odours. If concentrations are sufficiently high, then the odours may be directly responsible for symptoms such as changing in patterns of breathing and vomiting. The WHO recommends that within non-industrial environments, the concentration of any odourous compounds should not exceed its 50th percentile detection threshold. In practice, odours often comprise a mixture of VOCs. This makes the assessment of overall odour levels difficult. As a rule of thumb, it is usually the case that the overall odour intensity of a mixture of VOCs will be lower than the sum of the individual odour intensities of its individual components.

The issue of odours and the symptoms caused by odours is not the only factor associated with VOCs within the indoor in dwellings: within industrial buildings, the concentration of an individual VOC is likely to be high, and therefore it is a straightforward matter to determine whether the concentration will have an adverse effect on the health of occupants within the prevailing periods of exposure. Within a dwelling, the concentration of each VOC will be well below that at which it would be expected to have a noticeable adverse effect on occupants. Effects may arise from the combination of the range of VOCs present at low concentrations. Possible symptoms include sneezing, coughing and conjunctivitis. In extreme cases, skin swelling, rashes and breathing problems may result, but the cumulative concentrations of VOCs causing these should be very rare within the dwellings.

The main problem is that of representing the overall effect of the load and variety of VOCs within a given indoor environment. Accurate chemical analysis of indoor air for all VOCs would be the logical solution. However, this would be difficult to achieve (not to mention expensive) and it is probable that there would be a lack of repeatability from a set of samples taken from within one building alone. Even if the results were repeatable, the effects of synergism between certain compounds would be an extra complication. The setting of guideline concentrations for all VOCs would be of little value, and even if such guidelines were set, it would be impractical to try and enforce them.

A more realistic approach is required. There are two accepted options. The first is to set a scale of exposure-range classification for total VOC (abbreviated to TVOC) concentration in comparison to a calibration standard, usually toluene. The TVOC concentration excludes formaldehyde and any VOCs that are known to be carcinogenic. A comfort range of less than 200 $\mu g/m^3$ is suggested by both reference 29 and reference 30, whilst according to reference 29, discomfort due to VOC would not be anticipated until concentrations of between 3000 and 25,000 $\mu g/m^3$ were reached. Reference 30 suggests a range of concentration values above which certain adverse effects might be expected to be observed. These are summarised in Figure 2.5.

The second approach is to analyse the air and to determine the total concentration of VOC of the compounds within each of seven notional categories, rather than carrying out a comprehensive chemical analysis. The suggested target guideline for the TVOC is 300 $\mu g/m^3$, and in addition it is recommended that no single compound should exceed 50% of the target for its category or 10% of the TVOC concentration.

The first approach is the most readily applicable. Indeed, most air quality sensors are based on the blanket measurement of VOC with no discrimination between compounds. The second approach still involves more detailed chemical analysis. The practical results of the two approaches are, however, comparable.

In addition to considering the effects of TVOC, there are certain VOCs that have been shown to be carcinogenic and/or genotoxic. These include benzene, chloroform, tetrachloromethane, 1,2-dichloroethane and trichloroethylene. Others, such as styrene and 1,2-dichloroethane, may cause congenital abnormalities or diseases as a result of their mutagenic properties. Most people will only be exposed to very low concentrations of these particular compounds, and it is, therefore, very unlikely that any

TVOC concentration (mg/m^3)	Likely effects
<0.2	None
0.2–3.0	Irritation or discomfort if exposures interact
3.0–25.0	Irritation and headache if exposures interact
>25.0	Neurotoxic effects other than headache

Figure 2.5 TVOC concentrations and effects (after Molhave[30])

adverse effects would be caused. However, WHO takes the view that no safe threshold level should be assumed in such cases, and that the objective should be to keep concentrations as low as is possible.

2.11 Formaldehyde

Formaldehyde has been singled out because it has a certain notoriety amongst the compounds that could be grouped up together as VOCs.

Formaldehyde is the simplest of the aldehydes. It is colourless and highly chemically reactive to the point of being highly flammable. It has an extremely pungent smell. In the industrial field, it has has many uses; for example, as a preservative or sterilising agent. To many people of the authors' age group, the nasty smell of the pickled specimens in the biology laboratory was in fact formaldehyde.

Formaldehyde is a by-product of the human metabolism. The amounts involved are very small and pose no risk to health. There are many possible sources of formaldehyde within the environment, for example foods (it is sometimes used as a preservative), a wide range of household cleaning agents, paints and plastics. It is also contained within a number of materials that may be encountered in dwellings, used during construction and/or furnishing. Internal combustion engines without effective catalytic convertors, natural gas combustion, wood burning and tobacco smoking also provide sources of formaldehyde.

Whilst formaldehyde is readily soluble in water, it decomposes quickly and ingestion is not a realistic issue. In fact, exposure to formaldehyde may be regarded as mostly being due to airborne sources, and of these, combustion is by far the largest contributor. Airborne formaldehyde also decomposes, the main breakdown products being carbon monoxide and formic acid. Ambient air carries levels of formaldehyde which, not surprisingly, vary according to geographical location. In rural areas, concentrations may be as low as 0.2 parts per billion (ppb), whilst in urban locations, concentrations may be greater than 20 ppb when traffic flow is heavy. Skin absorption of formaldehyde is also possible.

Once absorbed into the body, formaldehyde is very easily broken down into formate (removed via urine) or carbon dioxide (removed via breathing). Some by-products are used in tissue building (happily, explaining this is beyond the scope of this book).

Direct contact with formaldehyde irritates human tissue, and the degree of sensitivity varies from subject to subject. The symptoms associated with direct contact include irritation of the eyes (with increased tear formation), nose and throat. These effects will be caused by airborne levels of between 0.4 and 3.0 ppm. There is some disagreement amongst the medical community as to whether formaldehyde exposure effects are more pronounced in asthmatic subjects.

If skin comes in contact with a strong formaldehyde solution, then irritation will take place. Prior to the use of dry ice, one of the more drastic remedies meted out by outpatients departments for veruccas was such a solution. (The author can testify to the irritative effects.)

Animal studies indicate that prolonged exposure to formaldehyde at concentrations in air above 6 ppm caused tumours in the nasal epithelium, or nose. With human subjects, such concentrations could not be tolerated due to the intense irritative effect. Therefore, it can reasonably be assumed that on the basis of present evidence there is little to suggest that the concentrations in indoor formaldehyde, likely to be observed, will be associated with any significant carcinogenic risk. However, the American Environmental Protection Agency (EPA) has stated that, based on the evidence available, formaldehyde is probably a human carcinogen.

Some problems arise as a result of the choice of building materials. The classic case of this is the use of urea–formaldehyde foam as an insulation material for wall cavities. Formaldehyde can out gas from the insulation for many years. Relatively low concentrations have been associated with irritated eyes and respiratory problems. In addition to this, there is some medical evidence which suggests that formaldehyde may be associated with increased risk of cancer, although it must be admitted that the evidence is far from conclusive. It should also be noted that formaldehyde is also a by-product of gas combustion and can, therefore, be present within the property even when foam insulation has not been used. In the USA, the use of formaldehyde has been perceived as such a big health risk; there are very large programmes instituted to remove formaldehyde from wall cavities and replace it with safe alternatives, at some considerable expense. Whilst safer alternatives have been found for formaldehyde as a component of wall cavity insulation, it is still the case that urea–formaldehyde-based glues are used in the production of the grades of wood-fibre boards used to make modern furniture. This means that sources of formaldehyde may in fact be unwittingly brought into the dwelling by the householders.

2.12 Subsoil gases

2.12.1 *Radon gas*

Radon occurs naturally as a by-product of radioactive decay of the elements, namely uranium and radium. These elements are found in all soils and rocks. Usually, soils will contain about 3 ppm of uranium. However, in some locations, particularly where the soil is located on top of granite or shale, concentrations as high as 10 ppm might be found. In some cases, sandstone and limestone deposits might give a similar effect. Radon is both odourless and colourless. The gas permeates its way through porous ground layers, and then it is emitted at the ground surface.

The negative pressures within buildings caused by the so-called stack effect (refer to Chapter 3) cause radon and other gases to be drawn in via floors and other routes (such as service pipe penetrations). The building envelope will act as a collector, preventing wind currents from sweeping the radon gas away. Some stones used as building materials also contain radioactive isotopes, and hence radon will be released into any dwelling made of such affected materials unless internal surfaces are treated. In Hong Kong, it is now well known that there is a significant problem with high indoor radon concentrations. Hong Kong does not have access to any significant amount of limestone-based materials suitable for the manufacture of cements and concrete. However, there are good supplies of pulverised fly ash, and the pozzolanic properties of this material have made it popular as a substitute for normal mortars. What was not realised, until relatively recently, was that the pulverised fly ash was a highly enriched source of the offending radioactive isotopes. The Hong Kong authorities are currently engaged in a large programme of work which not only involves quantifying the magnitude of the problem within the total building stock, but also seeks to identify effective remedial measures. The latter constitutes a massive and complex exercise, given the widespread use of high-rise buildings for both commercial and residential purposes within the former Crown colony. However, within the UK, emissions from building materials are not as important source as the ground itself, and therefore action to control emission from the former is not normally required.

Radon is a gas, and therefore may be inhaled. There is, however, a subtle yet more dangerous means by which it can enter the human body. Of course, radon is radioactive in its own right. This means that once it has entered the building, it will decay by means of alpha particle emission into a wide range of radioactive decay products, amongst them *polonium*. These decay products, referred to as *radon progeny* (or *radon daughters* in older texts), are produced in the form of solid particles rather than gas molecules. These can attach themselves to airborne dust particles. If such dust particles are inhaled, then polonium and the other radon progeny will be deposited within the upper part of the respiratory tract during the course of normal breathing. Continuing radioactive decay will lead to the irradiation of lung tissue in this region. It is well accepted that a consequence of exposure to radioactive material is an increased risk of cancer. In the case of radon and its progeny, the hazard is specifically an increase in the risk of developing lung cancer.

Outdoor concentrations of radon are not dangerous. Radioactivity levels of the order of $20\,Bq/m^3$ are common. Problems only arise if radon is allowed to permeate into an enclosed space above the ground, which is usually a building. Since the radon and its decay products cannot be dispersed except by means of ventilation, then concentrations will invariably rise. As the stack effect is temperature-difference driven, radon diffusion rates vary according to season and even time of day. Spot measurements of indoor radon concentration are, therefore, of little use in determining whether a risk to health may exist within a given building. Measurements over a time period of months are, therefore, preferred.

The concentrations of radon within dwellings vary significantly, even within the same locality. *Ground permeability, radon concentration in soil gas,* and most importantly, *the number and size of routes into individual dwellings* are the three most significant determinants of indoor radon concentrations. Indoor radon concentrations in excess of 500 Bq/m^3 have been observed in high-risk areas: the absolute maximum observed was 8000 Bq/m^3. Within uranium mines, where the effects of radon gas emission on human health were first noted, concentrations may be much higher than this.

Due to the relatively recent recognition of excessive indoor radon concentrations as a possible health risk, evidence which permits the quantification of the effects of prolonged exposure to a given radon concentration is scarce. The long-term risks attributable to radon exposure are believed to be a 30% increase in the risk of lung cancer for exposures of between 140 and 400 Bq/m^3, and an 80% increase in risk for an exposure of 400 Bq/m^3, relative to the risk at an exposure level of 50 Bq/m^3.

When assessing the risk to health attributable to radon, the hazard should be assessed in conjunction with effects of smoking. The Swedish study mentioned above also showed that smoking and radon exposure combine synergistically with respect to increasing the risk of lung cancer: that is, the combined risk for smoking and radon exposure together is greater than the sum of the individual increases in risk. In the case of smoking combined with radon exposure, the combined risk is closer to being the *product* of the two individual risks. The risk to heavy smokers could be as high as six times greater than for non-smokers. This will be mirrored in the average risk within the population as a whole, as only a minority of the population are smokers.

The International Commission for Radiological Protection (ICRP)[31] recommends an action level of 20 mSv (the Seivert (Sv) is a quantity that represents the actual radiation effect on tissue) per annum for exposure situations that are already in existence, and an action level of 10 mSv per annum for new exposure situations. The annual total effective dose from indoor radon exposure within the UK is estimated to be approximately 22,000 mSv.

The risks associated with radon exposure can be translated into a figure representing the number of excess deaths caused by exposure. It has been estimated for a number of years that exposure to high concentrations of radon within the home leads to approximately 360 excess deaths per annum due to the increased risk of lung cancer. Berry *et al.*[32] estimated the current lifetime risks of premature death from lung cancer for a range of radon exposures. These are summarised in Figure 2.6. Based on these figures, it is

Average radon exposure level (Bq/m^3)	Lifetime risk of early death through lung cancer
20	0.3
100	1.5
200	3.0
400	6.0

Figure 2.6 Variation of lifetime risk of early death through lung cancer with average radon exposure level

estimated that exposure to radon is responsible for approximately 2500 deaths from lung cancer per year. However, more recent research indicates that the number of deaths attributable in some part to radon may be very much higher than was previously thought to be the case. When deaths resulting from synergistic effects of smoking and radon combined are considered, a more representative figure for deaths associated with radon may be as high as 11,000 per annum.

There is some discrepancy over what constitutes a safe indoor concentration of radon, as the assessment of the risk is essentially based on statistical analysis. In the UK the so-called action level is set at 200 Bq/m^3. It has been estimated that in areas such as Devon and Cornwall, with greater than 30% of dwellings may have indoor radiation levels in excess of the action level. Means of controlling radon gas concentrations within dwelling are described in Chapter 6. Without going into detail at the moment, it should be pointed out that radon control cannot be achieved by enhanced ventilation of the occupied space.

2.12.2 *Landfill gas*

Landfill gas is a blanket term which is used to describe the cocktail of gases that is emitted from landfill sites as a result of the decomposition of organic waste. The process of decomposition (or biodegradation) is a complex one, and the rates of gas production and the amounts of each gas depend on the age of the landfill, the precise nature of the waste tipped and the amount of moisture present. There are two distinct stages of biodegradation. The early stages of the decomposition are aerobic in nature, as the microbes within the landfill are able to make use of oxygen in the air trapped within the waste as it has been tipped and compacted. As the oxygen is depleted, the decomposition mechanism changes to an anaerobic regime. It is at this stage that production of methane increases.

Methane is a colourless and odourless gas which has a density which is approximately two-thirds that of air. It is an asphyxiating gas, as it displaces oxygen from the air mixture. However, its flammability presents the greater danger:

- Between 5% and 15% concentration in air by volume, methane forms an explosive mixture; 5% and 15% are referred to as the lower and upper explosive limits, respectively.
- Below 5%, there is insufficient methane to sustain combustion.
- Above 15%, there is not enough oxygen in the air mixture for combustion to take place.

2.13 Indoor temperatures

Temperatures will be affected by ventilation strategy. The more ventilation that is supplied, the more energy will be consumed in restoring indoor air temperatures to the desired level. In cases where heating systems are of insufficient capacity, excessive ventilation may not only be a cause of high-energy consumption, but may also result in low indoor temperatures (refer to Chapter 3 for a worked example of how a reduction in internal temperatures will lead to an increase in relative humidity). The maintenance of temperatures in keeping with requirements for good health and comfort is essential. The living conditions with any property, large or small, owner occupied or rented, will never be satisfactory to the occupants if the internal temperatures are inappropriate.

Within the housing stock of the UK, protracted conditions of excessively high temperatures are unlikely to be observed. Doubtlessly, this will change as houses are built to progressively higher standards of thermal insulation.

In any case, the provision of rapid ventilation (refer to Chapter 7, where Building Regulations are discussed in detail), will usually be sufficient to control high internal temperatures in rooms (such as kitchens, where very high temperatures are likely to occur as a result of cooking activities during the summer). Considering the dwelling as a whole, it is most likely that the peak temperature will occur during the middle of the day as is the case with any other type of building. The reason for this is solar gains and the diurnal temperature cycle. The peak is occurring at a time when many houses are in any case unoccupied, and thus little inconvenience or discomfort will be experienced by householders. Occasional instances of excessive indoor temperatures will be tolerated as one of the features of summer. Given the nature of the climate in the UK, one might suspect that complaints of high indoor temperatures might be somewhat muted. Window opening will doubtlessly be the main method of controlling indoor temperatures within a dwelling. In secured environments, external doors might be opened as well.

As indoor temperature and external temperature converge, progressively higher ventilation rates will be needed in order to effect temperature control. Ultimately, when they are equal, no benefit will be derived by ventilating the dwelling, and in any case, there will be no temperature-driven force to promote ventilation; in other words effects of the incident wind has to be relied upon (refer to Chapter 3). When both internal and external temperatures are high, the benefits of cooling by ventilation will in any case be very small.

In summary, summer temperatures are for a variety of reasons not as serious an issue within dwellings as they are within offices. From the energy conservation point of view, it is disturbing to note that at a time when natural ventilation seems to be successfully being promoted as an energy conscious alternative to air conditioning, there is evidence of a growth in market penetration of home air-conditioning systems. It is to be hoped that this wanton and unjustified practice can be discouraged if not actually proscribed within future legislation.

Example calculations for the control of indoor temperature by ventilation are given in Chapter 3.

Waking up in a cold house during a winter morning might be an unpleasant experience. However, the effect is transitory, as the central heating system brings the building back upto an acceptable temperature. Under some conditions, people within a given building may have a perception that either the temperature within that building is either too warm or too cold. The effect may be disagreeable to the occupants, but will not cause any health issues. The most serious issues linked are related to longer-term exposure to extremes of temperature. Any problems related to temperature as experienced by the occupants of houses almost certainly will be associated with low temperatures.

In order to understand the issues involved, some explanation must be made of the way that the human body maintains its core temperature. The human body seeks to maintain a thermal balance at a core temperature of 37°C. This is the optimum temperature for the body and its complex assembly of chemical and biochemical processes to function. The thermal balance is shown in simple form in Figure 2.7. It is a trade-off between metabolic heat production from the functions of the body, the amount of clothing worn by the subject in question and heat gains either to or from the external environment. Even at rest, the human body emits $46\,W/m^2$, whilst medium activity will produce almost four times as much heat. This heat must be dissipated.

Insulation inhibits the loss of heat from a building, and it is the same for the human body. In the latter case, the insulation is provided by clothing. The "*clo unit*" was devised during World War II by British scientists who were looking for a simple way to represent the effectiveness of clothing as a body insulator. Gagge *et al.*[33] defined the clo as the amount of thermal insulation required to keep a sedentary subject comfortable at 21°C. One clo unit corresponds to a clothing thermal resistance of $0.155\,m^2K/W$. Typical values are summarised in Figure 2.8.

The body has several mechanisms at its disposal to regulate the core temperature. Under conditions of net heat gain, the body must dispose

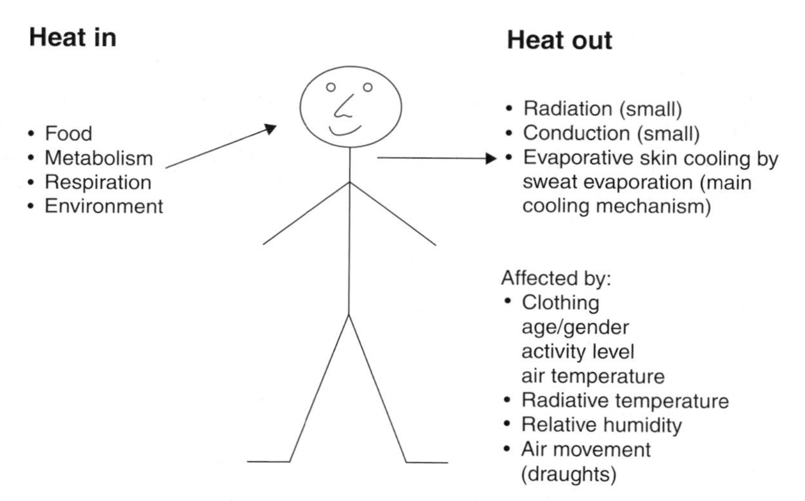

Figure 2.7 Basic body heat balance

Garment	Clo value
Naked	0
Light sleeveless dress + cotton underwear	0.2
Light trousers, vest, long sleeve shirt	0.7
As above plus jacket	0.9
Heavy three piece suit + "long johns"	1.5

Figure 2.8 Clo values (from reference[33])

off heat. This process is done by widening the blood vessels (also known as *vasodilatation*) of and also by sweating via pores in the skin. The greater the amount of heat to be removed, the greater the rate of sweating. Evaporation of sweat at the skin surface removes heat in the form of latent heat of evaporation. Under many conditions, sweating will be taking place with the subject being completely unaware. It is only when the sweat cannot evaporate at the rate at which it is being produced, then the thoroughly unpleasant feeling of wetness is experienced.

The issue of thermal comfort is a complex one. The perception of comfort by building occupants is a very individual matter. Calculations related to heat balance may be made, but ultimately the verdict on whether a particular subject feels thermally is heavily influenced by personal preference. The truth of this can be verified by observing the varied amounts of clothing worn by workers in an office building on a particular day. The main principle to understand is (to borrow an old politicians adage) that "it is impossible to please all the people all the time". Having said this, there are combinations of internal environmental conditions that will please more people than other types.

Much work has been published on the relationship between measured environmental parameters and subject perception of thermal comfort. The best known of these is that due to Fanger. His work is brought together in reference[34]. The author has found the book by McIntyre[35] to be an excellent source of information.

The main findings of Fanger's work have been codified into ISO Standard 7730 1984 (E)[36]. This document provides a means of determining the likely subject perception of thermal comfort in terms of the so-called predicted mean vote, or the PMV. This is a seven-point index that ranges from +3 (hot) through 0 (neutral, neither too hot nor too cold) and down to −3 (cold). PMV is determined from one of the most complex equations presented in the standard: it is not proposed to discuss this equation here. In order to avoid readers of the standard having to apply the equation, PMV values are presented in a convenient tabular form for a number of activity levels, clothing regimes and other environmental conditions. For the computer literates, a FORTRAN program code listing is also provided.

The PMV can be thought of as a "mean" perception of thermal comfort for a whole group of subjects. What it does not do is give any clue as to the likely spread of perceptions around the "mean". ISO 7730 also presents a means of determining the predicted percentage of dissatisfied, or PPD, which is related to the PMV by the simple equation:

$$\mathrm{PPD} = 100 - 95 \exp\{-(0.03353\,\mathrm{PMV}^4 + 0.2179\,\mathrm{PMV}^2)\}. \qquad (2.1)$$

Results for a range of PMV values, together with the distribution of voting, is shown in Figure 2.9.

In Annex A of ISO 7730, information is provided regarding the recommended comfort requirements. Interestingly, this annex is specifically and uniquely labelled because it is not forming the part of the standard. Annex A recommends that PPD should be kept below 10%, corresponding to a PMV of between −0.5 and +0.5. It also makes reference to the influence of asymmetry of heating or cooling on subjects, as might result, for example, from cold draughts on the neck or a cold floor. Annex A gives recommended comfort conditions for light mainly sedentary conditions during both winter and summer conditions. These are presented in tabular form for the purpose of comparison in Figure 2.10.

Critical assessments of the implications of Fanger's methodology are quite common. Understanding of the issues in minute detail is not necessary

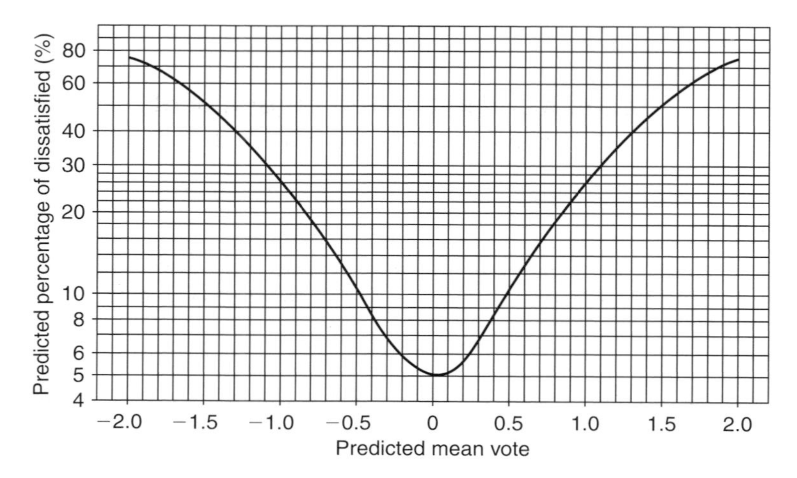

Figure 2.9 Predicted percentage of dissatisfied (PPD) as a function of predicted mean vote (PMV)

within the context of this book. Suffice it to say that the work of Fanger is not totally accepted by the building services profession despite its use within existing design criteria. In any case, the work has usually been targeted at commercial buildings rather than dwellings. It is true to say that of all the building types, the type for which understanding is least developed is for housing. This may well be due to the small groups of subjects in each dwelling, which makes the statistical validity of ISO 7730 highly questionable. In a situation where there is a need to achieve greater energy savings in dwellings, the issue of the determination of acceptable indoor air temperatures will, doubtlessly, come to the forefront. In many studies of the influence of energy-saving measures upon fuel consumption, it has been quite apparent that a proportion of dwelling occupants have effectively taken part of the predicted energy savings in the guise of increased indoor air temperatures. This is clearly an undesirable outcome.

From the viewpoint of simple design temperature considerations, it is probably far more useful for us to assess the suitability of temperatures within the dwellings. It is generally accepted that within the temperature range of 18–24°C, most people will feel comfortable. If this temperature range is not achieved, minor deficiencies are likely to result in discomfort rather than actual medical risks. It should be remembered that the human bodies have the capacity to acclimatise. At temperatures below 16°C, there may be actual health risks. Resistance to infections, such as the common cold and influenza, is decreased. This is especially true if low temperatures are experienced in conjunction with either high or low levels of humidity. Other problems may be caused, for example exercise-induced asthma in children and decreased resistance to infections of the bronchial epithelium. Heat losses via exhalation will be greater during conditions of low

Annex A

Recommended comfort requirements

(This annex does not form part of the standard.)

A.1 General

In this annex thermal comfort requirements are recommended for spaces for human occupancy. It is recommended that the PPD be lower than 10%. This corresponds (see figure 1) to the following criteria for the PMV:

$-0.5 < \text{PMV} < +0.5$

Corresponding comfort limits for the operative temperature may be found from the PMV index as described in clause 3. As an example comfort limits for the operative temperature are given in figure 2 as a function of activity and clothing.

As another important example, comfort limits for the operative temperature are listed A.1.1 and A.1.2 for light, mainly sedentary activity (70 W/m^2 = 1.2 met). This activity is characteristic of many occupied spaces, for example offices, homes, etc.

Sub-clause A.1.1 covers winter conditions where clothing of 1 clo = 0.155 m^2 · °C/W is assumed. Sub-clause A.1.2 covers summer conditions where clothing of 0.5 clo = 0.078 m^2 · °C/W is assumed.

The PMV and PPD indices express warm and cool discomfort for the body as a whole. But thermal dissatisfaction may also be caused by unwanted heating or cooling of one particular part of the body (local discomfort). This can be caused by an abnormally high vertical air temperature difference between head and ankles, by a too warm or cool floor, by air velocity being too high, or by a too high radiant temperature asymmetry. Limits for these factors are listed for light, mainly sedentary activity in A.1.1 and A.1.2. If these limits are met, less than 5% of the occupants are predicted to feel uncomfortable due to local heating or cooling of the body caused by each of the above mentioned factors. The percentages are not to be added.

The experimental data base concerning local discomfort is less complete than for the PMV and PPD indices. Sufficient information is thus not available to establish local comfort limits for higher activities than sedentary. But in general man seems to be less sensitive at higher activities.

If the environmental conditions are inside the comfort limits recommended in this annex, it can be estimated that more than 80% of the occupants find the thermal conditions acceptable.

A.1.1 Light, mainly sedentary activity during winter conditions (heating period)

The conditions are the following:

a) The operative temperature shall be between 20 and 24°C (i.e. 22 ± 2°C).
b) The vertical air temperature difference between 1.1 m and 0.1 m above floor (head and ankle level) shall be less than 3°C.
c) The surface temperature of the floor shall normally be between 19 and 26°C, but floor heating systems may be designed for 29°C.
d) The mean air velocity shall be less than 0.15 m/s.
e) The radiant temperature asymmetry from windows or other cold vertical surfaces shall be less than 10°C (in relation to a small vertical plane 0.6 m above the floor).
f) The radiant temperature asymmetry from a warm (heated) ceiling shall be less than 5°C (in relation to a small horizontal plane 0.6 m above the floor).

Figure 2.10

temperature. Even small reductions in body-core temperature and the temperature of body extremities will increase blood viscosity and hence tend to raise blood pressure in subjects of all ages. This increases the risk of stroke or heart attack due to the extra strain on the cardiovascular system. These findings are borne out by population studies. For example, it is now well established that for every 1°C drop in temperature below the average value during the winter period is associated with an increase in the death rate of about 8000. It has been reported by the Faculty of Health of the Royal College of Physicians of the UK[37] that the overall excess deaths (an "excess death" to a medical professional is an avoidable death) during the winter period are typically 40,000. Figures of between 29,000 and 40,000 have been previously estimated, depending on the severity of the weather. This level of excess winter deaths is not observed in other countries of comparable winter conditions but with higher standards of thermal insulation in dwellings, giving rise to the assertion that there are serious health problems associated with low internal temperatures within the UK housing stock. In Scotland, Wilkinson[38] in her review of the research evidence linking poor housing conditions and ill health, refers to data which suggests that excess deaths in Scotland during the winter lie in the range of 4000–7000 per year. Wilkinson also refers to the conflicting data relating to whether improvements in housing quality in Scotland could have led to a decrease in excess deaths during the winter.

The medical condition called *hypothermia* (meaning decrease in temperature) within the context of indoor cases has been well reported in the media since the large rises in domestic fuels which took place in the early 1970s. It is something of a surprise to find that the condition is in fact quite rare. Wilkinson[38] states that hypothermia accounts for about 1% of winter excess deaths within the UK.

Low temperatures are usually affected by the overall thermal insulation standard of the property in question and by the ability of the occupants to afford to use the amount of heat needed to maintain adequate temperatures. For more information on the incidence of low temperatures within the English housing stock, reference should be made to the section on the 1991 English housing condition survey.

A special mention should be made about the temperature requirements of elderly people. The thermoregulatory mechanisms of such people are often impaired. This means that they may in all probability not respond to low temperatures in the same way that a young healthy subject would. In the worst cases, core temperatures may drop to a dangerously low level very quickly, resulting in hypothermia, and this will lead to death if not detected and treated. For this reason, low temperatures within dwellings occupied by the elderly are likely to pose much more of a health risk than for younger subjects. In sheltered accommodation and nursing homes, it is relatively straightforward to control internal temperatures to a suitable level. The real problems are likely to arise in cases where elderly people are living in the dwellings which they either cannot afford to heat or do not understand how to heat in a cost-effective manner, for example keeping one room warm. The difficulties may be compounded if the subjects cannot afford to make repairs or refurbishments to the property that would reduce heat losses. Successive governments have attempted to deal with this problem, and it has to be said that some improvements have been made.

2.13.1 *Perception of air quality*

Humans have two senses that contribute to the overall perception of air quality. The general chemical sense is located all over the mucous membranes of the nose and also in the eyes. The olfactory sense is located in the nasal cavity. Both the senses are sensitive to hundreds of thousands of individual odorous substances in the air. The perception of air quality is a combination of the response of the two senses.

The perception of air quality is a complex subject. However, it can be said that of a given number of people exposed to the same mixture of odorous air pollutants, there will be a proportion who will be satisfied with the air quality on the basis of their perception, whilst the rest of the group will perceive the air quality as unsatisfactory and will thus by implication be dissatisfied.

It is likely that the favourable perception of indoor air quality will improve with time as people tend to become acclimatised. Acclimatisation is one of the several problems encountered when attempting to quantify the acceptability of indoor air within a space. A commonly used and probably reliable means of assessing perceived air quality is to determine the proportion of people who are dissatisfied just after they enter the space in question. Once again, there is an issue of practicality involved. It would be highly time consuming to carry out assessments of perceived air quality in all occupied spaces with a view to determine its degree of acceptability. What is needed is some method of being able to predict the likely degree of indoor air acceptability, given information about occupancy and pollutant production and from this decide on a ventilation rate that would be appropriate for the control of odour in a particular set of circumstances.

The ideal situation would be to be able to work out an index of occupant's satisfaction with their indoor environment, given knowledge of the concentrations of the pollutants within the indoor air. In practice, this is likely to be almost impossible for several reasons. The number of chemical compounds contributing to the overall occupants perception of odour is likely to be large (refer back to Section 2.10), so large in fact, that chemical analysis of even a few air samples would be a quite massive task. The equipment needed would not be portable, and hence it would not be feasible to carry out the chemical analysis on a real-time basis. The precise number of pollutants is likely to vary quite substantially from case to case, so that an "*evidence-base*" applicable to a large number of cases would be almost impossible to build up. As a further complication, some of the compounds that would show up as a result of chemical analysis would not in point of fact been present at the time that the air sample was taken. In fact they would be the decay products of one or more of the original VOCs that were present at the actual time of sampling, resulting from the delay time between the air sample being taken and the analysis taking place. Given the complexity of the chemical reactions taking place, it would be very difficult to relate the composition of the analysed sample to the collected sample.

Much of the contemporary research into the perception of air quality by occupants has been carried out by Fanger, who is of course the same researcher whose work on thermal comfort has made such a large contribution to current standards. His whole approach to air quality has been to try to characterise perception with respect to the body pollution generated, and relate perception to an odour standard. Fanger[39] has defined two important air quality parameters in his work:

1. The *olf* is the rate of body pollution production attributable to the average adult office worker in the sedentary state who feels thermally neutral. This means that the olf can be used as a measure of indoor air pollutant production, effectively odour intensities are expressed as multiples of the production rate of a standard person.
2. The second parameter is the *decipol*. This is defined as the perceived air quality in a space which has a pollution source strength of 1 olf and is ventilated by 10 l/s of fresh air; in other words, 1 olf/(l/s) is equivalent to 10 decipols.

Figure 2.11 shows the percentage dissatisfaction caused by a one olf pollutant source for a range of ventilation rates. Fanger defines three levels of perceived air quality, and relates these to decipol values. These are given in Figure 2.12, and are also shown in Figure 2.13.

The principal of the olf can be extended to encompass the emission of pollutants that are not bioeffluents; that is, not emitted by the human body. By means of the subjective assessment of the perception of a range of non-metabolic pollutants against standard smells (achieved by means of panels of trained assessors), Fanger has attempted to quantify the influence of all pollutant in terms of a total olf value for a given space. Knowing this total, a ventilation rate could be selected to give the decipol value appropriate to the chosen indoor air quality class. Olf levels for different types of building are listed in Figure 2.14. TVOC figures are also given for purpose of comparison. It is interesting to note that a figure for olf/m^2 are not quoted for the case of dwellings. It is suggested that this is a reflection of the diversity of sources within dwellings. Of course, it is also true that long-term acclimatisation is most probable with respect to dwellings. The smell of ones' own house is seldom noticed unless one has been away for a period of several days or more.

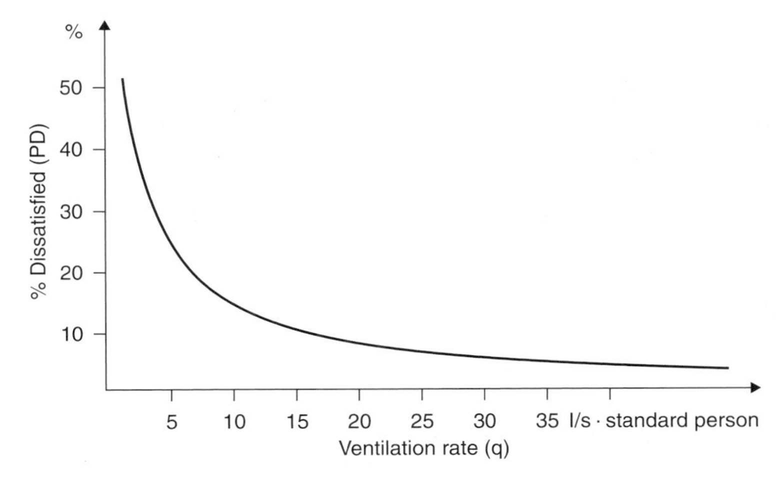

Figure 2.11 Dissatisfaction caused by a standard person (one olf) at different ventilation rates. The curve is based on European studies where 168 subjects judged air polluted by bioeffluents from more than one thousand sedentary men and women (18). Similar studies in North America (19) and Japan (20) by other research groups show close agreement with the present European data. The curve is given by these equations:

$$PD = 395 \cdot \exp(-1.83 \cdot q^{0.25}) \quad \text{for } q \geqslant 0.32 \text{ l/s} \cdot \text{olf}$$
$$PD = 100 \quad \text{for } q < 0.32 \text{ l/s} \cdot \text{olf}$$

Quality level (category)	Perceived air quality		Required ventilation rate* l/s-olf
	% dissatisfied	decipol	
A	10	0.6	16
B	20	1.4	7
C	30	2.5	4

* The ventilation rates given are examples referring exclusively to perceived air quality. They apply only to clean outdoor air and a ventilation effectiveness of one.

Figure 2.12 Three levels of perceived indoor air quality (examples)

If the work of Fanger in the area of thermal comfort has been high profile and somewhat controversial, then his research into air quality has also been much more so. Grave doubts have been expressed, both about the high air change rates implied by the olf method and also about the scientific rationale behind the method itself. The research has been in progress for more than 12 years at the time of writing. Between 1994 and 1997, concerted attempts were made to have the olf method incorporated into European legislation, in the form of a pre-adaptive norm. This norm would have become binding on all European Union (EU) members, who had two-thirds majority vote in favour, and had been obtained amongst

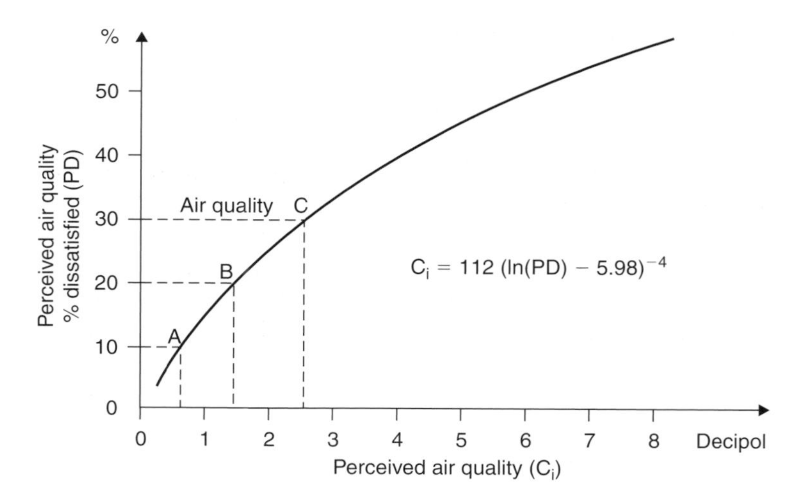

Figure 2.13 The relation between perceived air quality expressed by the percentage of dissatisfied and expressed in decipol (18). The three indoor air quality levels, categories A, B and C are shown.

	Sensory pollution load olf/(m^2 floor)		Chemical pollution load TVOC μg/s-(m^2 floor)	
	Mean	Range	Mean	Range
Existing buildings				
Offices[1]	0.3	0.02–0.95	–[6]	
Schools (class rooms)[2]	0.3	0.12–0.54	–[6]	
Kindergartens[3]	0.4	0.20–0.74	–[6]	
Assembly halls[4]	0.5	0.13–1.32	–[6]	
Dwellings[5]			0.2	0.1–0.3
Low-polluting buildings (target values)		0.05–0.1	–[6]	

1) Data for 24 mechanically ventilated office bulidings (30,31)
2) Data for 6 mechanically ventilated schools (28)
3) Data for 9 mechanically ventilated kindergartens (29)
4) Data for 5 mechanically ventilated assembly halls (30)
5) Data for 3 naturally ventilated dwellings (32)
6) Data not yet available

Figure 2.14 Pollution load caused by the building, including furnishing, carpets and ventilation system

the committee members. After many delays, the vote was taken in the autumn of 1997, and the requisite majority was not achieved. The future of the olf method within the EU legislative framework is not clear, but it would be most unwise to assume that it will be forgotten.

2.13.2 *Effectiveness of ventilation*

This section is purposely presented here rather than in Chapter 3, to indicate that it is a warning.

Even if the ventilation rate within a space may be shown to be adequate by calculation, it may prove to be the case that the situation might not be as good as in reality. All areas of the building might not be uniformly ventilated. To the occupant, what really matters is the quality of air within the so-called breathing zone, rather than in all the corners and little used spaces of the building. If the distribution of pollutants and hence air quality is not uniform, then it can be assumed that the rate of ventilation within the building is also not uniform. This is called *inhomogeneity*.

The measure of inhomogeneity of ventilation is the parameter known as *ventilation effectiveness*. This is given by

$$\varepsilon_v = \frac{C_e}{C_b} \tag{2.2}$$

where ε_v is the ventilation efficiency, and C_e and C_b are the pollutant concentrations in the exhaust ventilation air stream and the breathing zone, respectively. If the value of ε_v is >1, then this means that the concentration of pollutant in the breathing zone is lower than in the exhaust air stream. This means that there is an opportunity for the ventilation rate in the breathing zone to be reduced without compromising the air quality. However, if the concentration of pollutant is higher in the breathing zone than in the exhaust air stream, then the value of ε_v is <1, and this implies that the ventilation rate needs to be increased.

Ventilation effectiveness depends on the pattern of air distribution and the location of the pollution emission sources within the space in question. It may then be quite possible that the measured ventilation efficiency may differ from one pollutant to the next, since the locations of the emission sources may be different. A commonly used method of estimating ventilation effectiveness is used by Skaaret and Mathiesen.[40] The space under consideration is divided into two subspaces:

- the first subspace is that into which air is supplied,
- the second the rest of the main space.

When standard mixing ventilation strategies are considered, in which incoming supply air is mixed with the main body of the air in the space and extraction of air undertaken from this main body, then the optimum condition for air quality is when the two spaces merge into one. On the other hand, when a displacement regime is considered, the supply zone is occupied by people, and the exhaust zone is above the supply zone. This regime is the most effective when the amount of mixing between the two spaces is kept to a minimum.

The ventilation effectiveness within a space is also dependent on the siting of air terminals, air-flow rate and the temperature of the supply air. It is possible to determine ventilation effectiveness either by experimental measurement or by numerical simulation. In the absence of the time or inclination to perform either determination, the typical values presented by Skaret and Mathiesen may be used instead, as shown in Figure 3.9. These values assume even distribution of pollutant sources within the space in question, and therefore cannot represent the effects of different source distribution regimes.

In conclusion, great care must be taken to ensure that not only the quantitative ventilation requirement is known, but also the nature of the ventilation requirement is fully understood. If this is not the case, then indoor air pollutants will not be appropriately dealt with. In the subsequent chapters, several practical examples relating to ventilation efficiency will be encountered.

References

1 British Standards Organisation: BS5250. *Code for Control of Condensation in Buildings*. BSI, 1989; p. 55.
2 ibid, p. 5.
3 MJ Leupen. *Discussion in Indoor Climate 1979*. Danish Building Research Institution, 1979; pp. 124–126.
4 CA Hunter, C Sanders. English House Condition Survey 1991. HMSO.
5 G Raw, RM Hamilton. *Building Regulations and Health*. Construction Research Communications, London, 1995.
6 I Williamson, C Martin, C MacGill, R Monie, A Fennerty. Damp housing and asthma: a case control study. *Thorax* 1997, 52, 229–234.
7 J Evans, S Hyndman, S Stewart-Brown, D Smith, S Petersen. An epidemiological study of the impact of damp housing on adult health. Health Services Research Unit, University of Oxford, 1997.
8 C Martin, S Platt, S Hunt. Damp housing, mould growth and symptomatic health state. *British Medical Journal* 1987, 294, 1125–1127.
9 WA Croft, BB Jarvia, CS Yatawara. Airborne outbreak of trichothecene toxicosis. *Atmospheric Environment* 1986, 20, 549–552.
10 R Voorhorst, FTM Spieksma, H Verekamp, MJ Leupen, AW Lykema. The house dust mite and the allergen it produces. *Journal of Allergy* 1967, 39, 325–339.
11 C Taylor. *Fight the Mite: a Practical Guide to Understanding House Dust Mite Allergy*. The Wheatsheaf Press, 1992.
12 TAE Platt-Mills, MD Chapman. Dust mites: immunology, allergic disease and environmental control. *Journal of Allergy and Clinical Immunology* 1987, 80, 755–775.
13 S Owens, M Morganstern, J Hepworth, A Woodcock. Control of house dust mite antigen in bedding. *The Lancet* 1990, 335, 396–397.

14 H Harving, LG Hansen, J Korsgaard, PA Neilsen, OF Olsen, J Romer, UG Svendsen, O Osterballe. House dust mite allergy and anti mite measures in the indoor environment. *Allergy* 1991, 46, 33–38.
15 H Harving, J Korsgaard, R Dahl. House dust mites and associated environmental conditions in Danish homes. *Allergy* 1993, 48, 46–49.
16 M Winkmann, G Emenius, A Egmar, G Axelsson, G Pershagen. Reduced mite allergen levels in dwellings with mechanical exhaust and supply ventilation. *Clinical and Experimental Allergy* 1994, 24, 109–114.
17 DA McIntyre. The Southampton survey on asthma and ventilation: humidity measurements during winter. *16th AIVC Conference*, Warwick, UK, 1995.
18 AM Fletcher, CAC Pickering, A Custovic, J Simpson, J Kennaugh, A Woodcock. Reduction in humidity as a method of controlling mites and mite allergens: the use of mechanical ventilation in British domestic dwellings. *Clinical and Experimental Allergy* 1996, 26, 1051–1056.
19 A Custovic, SCO Taggart, JH Kennaugh, A Woodcock. Portable dehumidifiers in the control of house dust mites and mite allergens. *Clinical and Experimental Allergy* 1995, 25, 312–316.
20 IM Samet, MC Marbury, JD Spengler. Health effects and source of indoor air pollution, Part 1. *American Review of Respiratory Disease* 1987, 136, 1486–1508.
21 The American Lung Association (ALA), The Environmental Protection Agency (EPA), The Consumer Product Safety Commission (CPSC) and The American Medical Association (AMA) (all co-sponsors): Indoor Air Pollution: An Introduction for Health Professionals. US Government Printing Office Publication No. 1994-523-217/81322, 1994.
22 The American Thoracic Society. Report of the ATS Workshop on Environmental Controls and Lung Disease, Santa Fe, New Mexico, 24–26 March 1988. *American Review of Respiratory Disease* 1990, 142, 915–939.
23 Health and Safety Executive, EH40, The Stationary Office, 2002.
24 B Joynt, S Wu. *Nitrogen Oxides Emissions Standards for Domestic Gas Appliances – Background Study*. Department of the Environment and Heritage, Australian Government, February 2002. (Also downloadable from http://www.deh.gov.au/atmosphere/airquality/residential/)
25 GJ Raw, SKD Coward. Exposure to nitrogen dioxide in homes in the UK: a pilot study. *Unhealthy Housing: The Public Health Response*, University of Warwick, 18–20 December 1991.
26 CR Wiech and G Raw. The effect of mechanical ventilation on indoor nitrogen dioxide levels. *Proceedings of Indoor air '96, Nagoya*, Japan, Vol. 2. 1996; pp. 123–128.
27 DW171. *Standard for Kitchen Ventilation Systems*. HVCA, 1999.
28 http://www.airquality.co.uk/archive/standards.php#std
29 *Indoor Air Quality Guidelines for Europe*, 2nd Edition. World Health Organisation Regional Office for Europe, 2000. (Available on CD, and also downloadable from http://www.euro.who.int/eprise/main/who/progs/aiq/activities/200206201)
30 L Molhave. Volatile organic compounds, indoor air quality and health. *5th International Conference on Indoor Air Quality and Climate*, Toronto, Vol. 5. 1990; pp. 15–33.
31 *ICRP Publication 65: Protection Against Radon-222 at Home and at Work*, 65 Annals of the ICRP Volume 23/2, International Commission on Radiological Protection. ISBN: 0-08-042475-9, 1994.

32 RW Berry, VM Brown, SK Coward, DR Crump, M Gavin, CP Grimes, DF Higham, AV hull, CA Hunter, IG Jeffrey, RG Lea, JW Llewellyn, G Raw. *Indoor Air Quality in Homes*. The BRE indoor Environment Study. Construction Research Communications, 1996.
33 AP Gagge, AC Burton, HC Bazett. A practical system of units for the description of the heat exchange of man with his thermal environment. Science NY, Vol. 94, pp. 428–430, 1941.
34 PO Fanger. Thermal comfort. McGraw Hill, New York, 1972.
35 DA McIntyre. Indoor climate. Applied Science Publishers, England, 1980. ISBN 0-85334-868-5.
36 ISO 7730: Moderate thermal environments – determination of the PMV and PPD indices and specification of the conditions for thermal comfort, 2nd edition. International Standards Organisation, Geneva, 1984.
37 Royal College of Physicians Media release, December 2003 (Downloadable from http://www.fphm.org.uk/publications/Media_Releases/PDF_files/2003/cold_snap_causes_unnecessary_deaths_dec03.pdf).
38 D Wilkinson. Poor housing and ill health – a summary of research evidence. Scottish Office Central Research Unit, 1999.
39 PO Fanger. A comfort equation for indoor air quality and ventilation. *Proceedings of Healthy Buildings '88*, Stockholm, Vol. 1. 1988; pp. 39–51.
40 E Skaaret, HM Mathiesen (1982) *Ventilation Efficiency*. Environment Int. 8: 473–481.

3

Prediction Techniques

Building simulation can be quite readily achieved with modern computers and the powerful software that can be run in such machines. A wide range of aspects of building performance can be simulated, ranging from energy consumption through to the effectiveness of internal lighting schemes and the influence of daylight. Previously, many of these predictions are carried out using manual calculation techniques. In some cases, the calculations required were far too complex for manual purposes. The ever-increasing power of the desktop computer makes such tasks easy to complete in a very short period of time.

One consequence of current developments that might be construed as a retrograde step is that calculation power is being achieved at the expense of usability, and user understanding of the principles and processes that are being used to carry out the calculations. There are people who would argue that the television viewers do not need to understand how a television works, or that the owners of a home personal computer (PC) do not have to possess any knowledge of the principles of microelectronics in order to be able to use their machine. This line of argument may have great validity in these cases.

However, in the case of building simulation software, there is some degree of risk which is associated with the unreasoned production of simulation results. Some of the suites of thermal prediction software that are presently available require an appreciable amount of user training, and are beyond the expertise of many potential users. Without an understanding of the basic principles behind the simulation technique in use, the danger is that the unquestioning user might not be able to fully understand the significance of results obtained, or worse still will not be able to spot when a simulation is producing spurious results as a consequence of inaccurate or inappropriate input of building data. The old computer programmers' adage of "garbage in, garbage out" is clearly applicable in such circumstances.

With some software packages, it might be very difficult even for an experienced user to retrace the data entry process and spot the offending input error. In some cases, perhaps surprisingly to some readers, versions of simulation software reach the end user with errors in the programming code. The consequence of such "bugs" might be the much irritating thing due to the inability to access certain features of the software, or else a tendency for the software to inexplicably "crash" at an inopportune moment. Such problems should be rapidly identified by the users of the software. As a result of this the software vendor will be forced to issue a modified, "bug-free" version at the earliest possible opportunity.

In practice, the programming errors that present the biggest risk to the unquestioning user are those within the calculation algorithms of the software. It will often be the case that such errors will not cause any problems other than erroneous results, the magnitude of which may vary considerably. The user who understands the principles behind the calculation algorithms stands a chance of realising that the results obtained are wrong. If the user's expertise goes no further than accepting the results at face value, then they may be in some difficulty at a later date. The author, who also lectures to undergraduates on building service-related matters, notes with some considerable concern the ease at which students from a computer literate generation take to the use of simulation tools, coupled with the unquestioning faith that they place in the answers generated. If this attitude becomes prevalent in design offices, then there are going to be some major design disasters in the very near future; in other words, some have not already been taken place.

In the final analysis, the use of highly involved methods is not justified in many cases. Fortunately, these are still available for the casual user, as a wide range of calculation tools that will yield usable results without the need for great computing power or extensive training. With a little knowledge, many practitioners can produce useful results and assess their significance.

This chapter looks at the general principles behind a range of tools and procedures currently available to those interested in predicting ventilation requirements, air infiltration rates and interzonal air movements. The vast majority available are calculations or computations that can be carried out using nothing more than a pocket calculator or a computer spreadsheet. To assist the reader in understanding the procedures, calculations and worked solutions are provided throughout the chapter.

3.1 Assessment of ventilation requirements

The type of ventilation prediction technique used will depend on the objective that is to be met. For example, when assessing the likely energy consumption of a building, consideration must be given to the complexity of the problem being analysed and the precision required from the results produced. It would be unnecessary to make use of a sophisticated dynamic simulation, then only a simple estimate of steady-state ventilation heat losses for a typical winter day would suffice.

Similar considerations apply to the modelling and prediction of air infiltration and interzonal air movements (the latter being the movement of air between connected spaces within the same building). In many cases, detailed procedures are not required, and relatively simple tools can be used to produce fairly reliable results for the majority of the needs of the users. In a smaller number of cases, for example those where contaminant removal between rooms or parts of rooms is an issue, complex interzonal air-flow models will have to be used. It is true to say that air-flow modelling requirements within dwellings are not as complicated as those within large commercial and industrial buildings, where the size of individual zones might mean that a network of sub-zones with different ventilation and air-flow characteristics might exist within a given large space. For more information about large buildings, readers are recommended to look at the paper by Stymne *et al.*[1]

3.1.1 *Condensation control*

It should have been appreciated from previous chapters that within dwellings in the UK, the control of condensation is currently the most important objective of any credible domestic ventilation strategy. Therefore, it is the issue of the prediction of ventilation rates for condensation towards which the attention of the reader is first turned.

It will be remembered that the actual ventilation requirement for condensation control will not, in most circumstances, simply be the minimum ventilation rate that will prevent condensation physically running down the walls of the dwelling. As has been explained in Chapter 2, mould growth can occur, and for that matter house dust mites can thrive, under conditions where surface condensation is not taking place on a regular basis. The real danger is protracted periods of high mean relative humidities, typically in excess of a month. In general, mean indoor relative humidities must be kept below 70%, if mould growth is to be avoided. To control the breeding of house dust mites, the absolute moisture content of indoor air should be kept below 7 kg/kg dry air.

The existence of the conditions for mould growth can often arise as a function of high-moisture production. It is, therefore, logical that when attention is given to the issue of condensation control, the normal first step would be to look at the issue of ventilation rate. As will be seen later on, the internal air temperature within the space has an influence on relative humidity, since relative humidity is of course not merely a function of air moisture content but also of air temperature. Strictly speaking, it is not the case that ventilation alone is the answer to any given problem, although it will very often prove to be one of the most important factors. Insulation and heating are also the important factors in the control of condensation and mould growth. This should not be overlooked when assessing the problem in dwellings.

In order to stand a chance of avoiding mould growth, it will be necessary to estimate the magnitude of an appropriate ventilation rate for a given dwelling. There are two types of methods by which this can be done, namely by steady-state or non-steady-state methods. A steady-state calculation involves taking a set of appropriate data (e.g. a set of average daily environmental design parameters) in order to determine a mean result. As will be seen, required ventilation provision can be calculated with knowledge of internal air moisture content and temperature, together with the same data for outside air. The same procedure can be easily adapted to yield the likely internal air moisture content, given the chosen value of air change rate, internal and external temperatures, and external air moisture content.

A dynamic calculation procedure is intended to yield information about the transient variation of chosen parameters. For example, such a procedure might be used to predict the maximum internal relative humidities within the chosen cells in a dwelling, or else to yield daily profiles of internal air moisture contents. Whilst this data is undoubtedly useful in some circumstances, the associated increase in complexity of the input data required will often not justify the ends. Taken to the logical conclusion, the most reliable means of performing a dynamic ventilation simulation is probably to do it as part of an overall comprehensive simulation of the performance of the dwelling. In order to be able to achieve this, the amount of data input that will be required is considerable; so much in fact that not all of the necessary data items may be known with certainty. This puts the person performing the simulation in a position where they have to make educated guesses at the values of certain data items. Therefore, in seeking to use a methodology that reduces the uncertainty of the results obtained, approximations are made which lead to increased uncertainty as to the reliability of the results produced. This is a common problem in building simulations of all types.

Steady-state calculations are much easier to perform than non-steady-state ones, and in many cases this factor alone justifies their use. However, it must be realised that they also have their limitations. They can only be used for carrying out either snapshot assessments or else medium- to long-term average calculations, and will not give any indication of the danger of high-intermittent peaks of indoor air moisture content and/or relative humidity in the same way as a transient calculation. However, the long average figure is a very useful one, given that the most widely used criterion for the control of mould growth and condensation is that the average relative humidity should not exceed 70% over lengthy periods of time.

The following parameters are used in the ventilation rate calculation: external and internal air moisture content, and the rate of moisture production in the space. Moisture production can be a result of human occupation or activities, such as bathing and clothes washing.

It can be shown that the internal air moisture content (g_i) is given by the formula:

$$g_i = \frac{g_o + M}{(\rho VN)}, \tag{3.1}$$

where g_o is the moisture content of external air (kg/kg of dry air); g_i, the moisture content of internal air (kg/kg of dry air); M, the moisture generation rate in the space (kg/h); V, the volume of the space (m^3); and N, the air change rate (air changes per hour).

Example 3.1

Estimate the minimum air change rate required to avoid mould growth in a space of 100 m^3 volume, given the following data:

- External conditions: 95% relative humidity, 0°C.
- Moisture generation rate in the space: 2 kg/h.
- Required air temperature in the space: 20°C.

Solution:

The minimum air change rate is of interest in this case. The critical relative humidity is 70%. From the psychrometric chart, it can be seen that $g_o = 0.05$ g/kg and $g_i = 0.15$ g/kg. Rearrangement of Equation (3.1) gives:

$$N = \frac{M}{(g_i - g_o)\rho V}, \tag{3.2}$$

from which the minimum required air change rate in this example is:

$$\frac{2}{(0.15 - 0.05) \times 1.2 \times 100} = 0.167 \text{ air changes per hour.}$$

It is a common (not to mention prudent) practice to build in a safety margin when any type of calculation is made for design purposes. The same practice should ideally be applied to the steady-state air change rate calculation. The internal relative humidity could, for example, be set below the 70% threshold, say at 65%. The psychrometric chart is used to convert this to a corresponding moisture content of 0.12 g/kg dry air. Substituting this value of g_i into Example 3.2 gives:

$$N = \frac{2}{(0.12 - 0.05) \times 1.2 \times 100} = 0.238 \text{ air changes per hour.}$$

As might be expected, the consequence of seeking to maintain a lower relative humidity is an increase in the predicted air change rate. In this case, the required air change rate is more than doubled. The increase would, of course, have implications for energy consumption. However, as will be seen in Section 3.1.4 below, it is not true that the effect of increasing air change rate will always be a reduction in relative humidity.

Example 3.2

For the same room and conditions as given for Example 3.1, estimate the minimum air change rate required in order to inhibit the breeding of house dust mites.

Solution:

In this case, a moisture content of indoor air of 0.07 kg/kg dry air should be selected, since at or below this value, the breeding of house dust mites will be inhibited. This is rather lower than the values used in Example 3.1. Use of Equation (3.2) gives:

$$N = \frac{2}{(0.07 - 0.05) \times 1.2 \times 100} = 0.833 \text{ air changes per hour.}$$

This calculation serves to underline the fact that attempting to control house dust mites will require rather larger air change rates than would be necessary to prevent mould growth on surfaces. In practice, this will be difficult to achieve on a regularised, controllable basis by natural ventilation alone. Mechanical ventilation would probably be the only means by which higher air change rates could be reliably achieved, and the preferred strategy would be both the balanced supply and extract. This has indeed been used as a justification for the specification of mechanical ventilation with heat recovery, as will be discussed in Chapter 6.

It should be noted that Equations (3.1) and (3.2) are not specific to water vapour alone. They can be used to predict steady-state ventilation requirements for other chemically stable contaminants (i.e. those that do not decompose with time), such as carbon dioxide (CO_2) and carbon monoxide (CO). It is not proposed to give example calculations for all contaminants here, but suffice it to say that the air change rates for condensation and mould growth control are very much higher than for any of the other contaminants normally found in the air within dwellings. This means that, in practice, if the ventilation rate is successful in controlling mould growth, then there should in principle be no problems experienced with excessively high concentrations of the other contaminants.

The presence of such a problem would in all likelihood be indicative of a design problem, an equipment defect or an unusually high input of pollutant from building materials. For example, incidences of high CO concentrations within dwellings are more likely to occur as a result of inadequate fresh air supply to an appliance, a faulty or poorly serviced gas appliance, or even inappropriate or defective flue arrangements. In such cases, no change in ventilation strategy could reasonably be expected to provide a solution.

3.1.2 *Surface temperatures and surface condensation risk*

In practice, whilst the moisture content of the air mass is an important factor with respect to mould growth, condensation itself takes place on surfaces. Condensation occurs when the temperature of a particular surface is lower than the dewpoint temperature of the mass of air. In practice, this means that condensation will be a risk on the internal faces of exposed surfaces; in other words, on window glazing, metal window frames and on the inner surfaces of external walls. Therefore, in order to get a full picture of the potential risk, the thermal properties of the elements making up the building envelope must be taken into account, and the associated inner surface temperatures should be calculated in order to make sure that they do not fall below the dewpoint temperature.

The most convenient means of calculating the inner surface temperature of a structure is a truncated version of the overall procedure used to assess the risk of interstitial condensation. Essentially, the procedure consists of the following stages:

1. Select the internal and external temperatures, and relative humidities for the prediction.
2. Calculate the internal dewpoint temperature for the air within the dwelling. This can be done either by calculation, by tables or else by the use of psychrometric chart in Figure 3.1. It is not necessary to calculate the internal and external vapour pressure difference, as would be the case in an interstitial condensation risk assessment. This is because the driving force for water vapour flow across the structure has no bearing on the surface condensation risk.

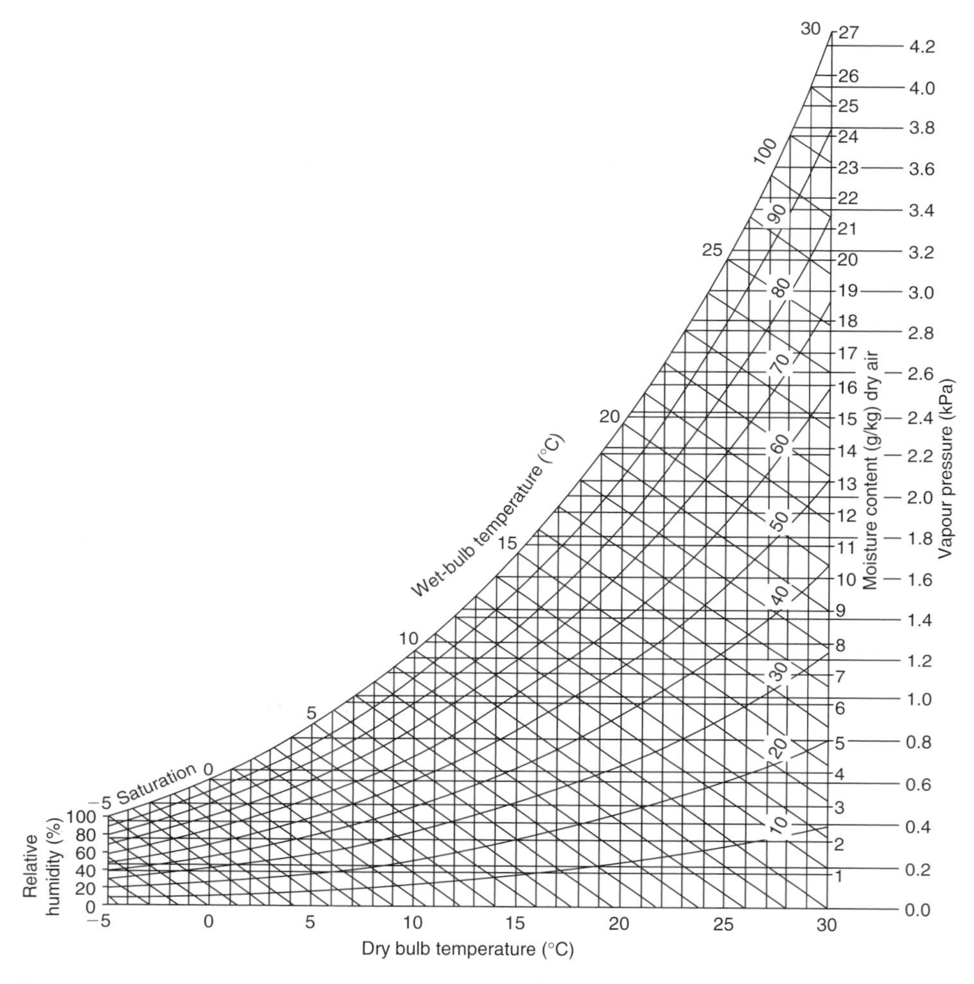

Figure 3.1 Psychrometric chart (*Source:* BS5250[2], no permission needed)

3. Determine the *U*-value for the whole structure. This can be done using thermal resistivity data from several sources. BS5250 contains suitable data. If an overall figure for a generic type of wall construction is selected, it will still be necessary for the thermal resistance of the internal surface to be known. The selection of the value of this quantity should be done with some caution, as the surface resistance can have a large impact on the predicted inside wall surface temperature. A value of 0.12 m^2K/W fairly represents a vertical internal surface. As will be seen later in Section 3.1.3, this value is acceptable only for the bulk surface of walls. Special attention should be given to corners.

4. Determine the heat flux across the structure using the equation:

$$Q = \left[\frac{1}{\Sigma R}\right](T_i - T_o), \tag{3.3}$$

where Q is the heat flux (W/m^2); ΣR, the sum of the thermal resistances (m^{2}K/W) ($1/\Sigma R$) is therefore the U-value of the wall including the internal surface resistance and T_i and T_o are the internal and external surface temperatures, respectively (°C or K).

5. Determine the temperature of the inner surface using the equation:

$$T_s = T_i - QR_{si}, \tag{3.4}$$

where R_{si} is the internal surface resistance (m^{2}K/W).

6. Compare the calculated inner surface temperature with the dewpoint temperature. If the structural temperature is lower than the dewpoint temperature, then condensation will take place on the surface.

Example 3.3

The U-value of a wall is 0.6 W/m^{2}K. Calculate whether condensation takes place on the inside surface under the following environmental conditions:

- Internal condition: air temperature is 20°C and relative humidity is 60%.
- External condition: air temperature is 0°C and relative humidity is 90%.

It should be assumed that $R_{si} = 0.12$ m^{2}K/W.

Solution:

From the psychrometric chart, the dewpoint temperature is 13°C. Using Equation (3.3), the heat flux across the structure is given by:

$$Q = U(20 - 0) = \left(\frac{1}{1.78}\right)20 = 11.24\ \text{W/m}^2.$$

Hence from Equation (3.4) the surface temperature (T_s) is given by:

$$T_s = 20 - (11.24 \times 0.12) = 18.65°\text{C}.$$

This means that in this case no surface condensation would be expected on the wall.

In dwellings of modern construction within the UK, U-values are 0.45 W/m^{2}K or better, perhaps going as low as 0.25 W/m^{2}K. In general, therefore, the risk of condensation on internal wall surfaces has been significantly reduced. Problems on such surfaces are only likely to occur when buildings are inadequately heated, or when ventilation is grossly inadequate. Danger points for condensation will still exist. These will be in regions where the thermal transmittance across the building envelope is higher than the main area of external wall. These are now likely to be window panes and frames, cold bridges, and internal corners.

The risk of condensation on window panes depends on the type of glazing. With single glazing, it is almost impossible to avoid the problem, whilst triple-glazed panes virtually eliminate surface condensation. The risk of condensation on window frames made of non-timber materials has been massively reduced by the inclusion of thermal breaking. Other forms of cold bridging are essentially design problems, notwithstanding that poor practice during construction may also cause difficulties. In certain types of construction, for example where steel beams are used, the risk might be quite high if appropriate and adequate insulation is not provided. The topic of cold bridging is a complex one, and beyond the scope of this book. For information on how to avoid the risk of cold bridging, the reader is recommended to consult BRE's Guide 'Thermal Insulation: Avoiding Risks'.[4]

Example 3.4

Calculate whether condensation would take place on:

(a) a single-glazed window pane or
(b) a double-glazed window pane.

for the same environmental conditions used for Example 3.3.

It should be assumed that the *U*-values for single- and double-glazed panes are 1.9 and 0.9 W/m^{2}K, respectively.

Solution:

Case (a): using, Equation (3.3) as before gives:

$$Q = U \times (20 - 0) = 1.573 \times 20 = 31.46\,W/m^2,$$

and hence from Equation (3.4) the temperature of the inside surface of the window pane is:

$$T_s = 20 - (31.46 \times 0.12) = 16.22°C.$$

This means that surface condensation would be expected on the inside surface.

Case (b): following the same procedure,

$$Q = 1.231 \times 20 = 24.62\,\text{W/m}^2,$$

from which

$$T_s = 20 - (24.62 \times 0.12) = 17.0°\text{C}.$$

Thus there is no condensation on the inner surface of the double-glazed window. This result is consistent with the real-life observation that properly installed double-glazed windows are less susceptible to surface condensation than single-glazed panes. However, they are not the total solution to the problem, since at low internal and external temperatures, coupled with high water vapour contents in internal air (as might be encountered in bedrooms during the night when the central heating is switched off), condensation will be unavoidable. When the seal between the two panes of glass in a double-glazed unit fails (the expected life span of such a unit is estimated at no more than 15 years) condensation will take place between the panes. When this happens, the only remedy is to replace the unit.

3.1.3 *The prediction of condensation risk in corners of rooms*

Examination of occurrences of mould growth within the dwellings shows that in the majority of cases, the most common location for persistent problems is at the inner surface of external corners. This real-life observation is entirely consistent with the hygrothermal regime, which is likely to prevail in such areas. There are two factors that combine to cause a greatly increased risk of condensation and mould growth in corners. These are shown in Figure 3.2. Firstly, the external corner of a building effectively acts as a cooling fin. This means that heat losses will be greater in the vicinity of the corner and hence the inside surface temperature will be reduced in comparison to the temperature of the rest of the internal walls.

In addition to this, the air-flow pattern at the inner surface of the corner will not be the same as over the open areas of internal wall. There is a stagnant pocket of air in the corner that allows a build-up of water vapour, hence resulting in a localised increase in relative humidity. As a side effect, it also increases the perceived internal surface thermal resistance, thus compensating in some small way for the greater heat loss caused by the fin effect, but nowhere near enough to prevent the problems.

The mathematical resolution of this matter is complex. The solution of the problem involves the computation of two-dimensional heat flows and so there is no simple, easy-to-apply equation that can be derived from the first principles. The current favoured treatment is the use of the empirical equation derived from laboratory measurements of the heat flow regime within corners. The results of the study suggest the form of

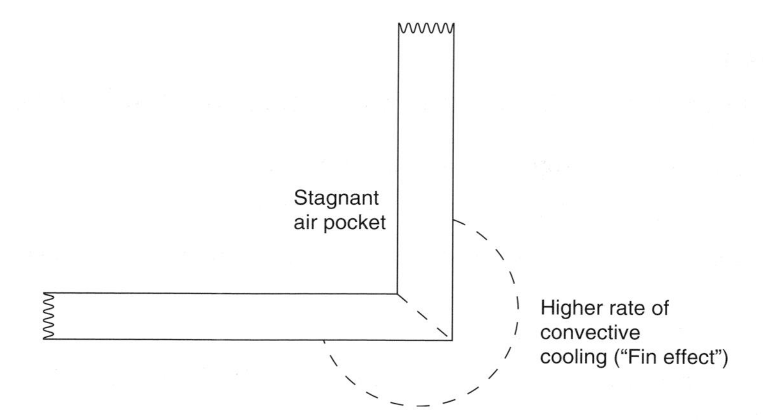

Figure 3.2 Factors influencing condensation in a corner

the empirical relationship as:

$$T_{\text{corner}} = \frac{T_i - 3R_{si}(T_i - T_o)}{3R_{si} \times 3R_{so}}, \tag{3.5}$$

where T_i is the internal air temperature (°C); T_o, the external air temperature (°C); T_{corner}, the internal corner temperature (°C); R_{si} and R_{so} are the internal and external surface thermal resistances, respectively (m^{2}K/W) and

$$R = (1/U) - (R_{si} + R_{so})$$

where U is the thermal transmittance of the external walls (W/m^{2}K). The author is compelled to admit that even after extensive consultation with colleagues, he is unable to find the original reference to this equation, which has formed part of his lecture notes for over 15 years!

The approximate nature of the relationship should be noted. However, it has been shown to give a fairly reliable answer in most cases. Note that because of its empirical nature, the relationship does not contain a term that directly reflects either the presence or the properties of stagnant air in the corner.

Example 3.5

Calculate the risk of condensation for both the main surface of the walls and also in the corner for the following cases:

- Internal conditions: temperature is 20°C and relative humidity is 65%.
- External conditions: temperature is −5°C and relative humidity is 95%.
- U-value of wall is 0.8 W/m^{2}K,
- R_{si} = 0.12 m^{2}K/W; R_{so} = 0.06 m^{2}K/W.

Solution:

From the psychrometric chart, the dewpoint temperature of the internal air is 16°C.

The heat flux through the wall is given by:

$$Q = 0.8 \times (20 - (-5)) = 0.8 \times 25 = 20\,\text{W/m}^2.$$

Using Equation (3.4), the surface temperature of the main wall area is given by:

$$T_{si} = T_i - 20 \times 0.12 = 20 - 2.4 = 17.6°\text{C}.$$

Using Equation (3.5), the surface temperature in the corner of the room is given by:

$$T_{corner} = 20 - \frac{3 \times 0.12[20 - (-5)]}{R + 3 \times 0.12}.$$

In this case, $R = (1/0.8) - (0.12 + 0.06) = 1.07$ and hence

$$\begin{aligned} T_{corner} &= 20 - \frac{0.36 \times 25}{1.07 + 0.36} \\ &= \frac{9}{1.43} = 6.29 \\ &= 20 - 6.29 \\ &= 13.71°\text{C}. \end{aligned}$$

Comparing the two results with the dewpoint temperature shows that whilst there is no problem with the wall surface itself, there is a condensation risk in the corner.

This is a problem that is relatively uncommon in modern houses with higher thermal insulation standards. As a result of better standards of insulation, even surface temperatures in the corners should be well above that corresponding to the dewpoint temperature. On the assumption that the property in question is adequately heated, the best remedy would be to increase the surface temperature in the corner by increasing the thermal resistivity of the wall. This would be achieved by the addition of extra insulation material. It is possible to calculate how much extra insulation would be needed in order to prevent condensation in the corner, using the following simple calculation procedure.

In practice, it may be difficult to apply extra insulation solely to corner elements. The preferred solution would probably be to locate insulation externally in the form of cladding. This has the added benefit of reducing the risk of interstitial condensation furthermore, providing the cladding is fitted satisfactorily, the rate of air infiltration through the building envelope should be significantly reduced. External cladding is a popular element of refurbishment packages for high-rise blocks of flats.

3.1.4 *The estimation of optimum ventilation rate*

The procedure previously described in Section 3.1.1 yields an answer that can be regarded as a minimum air change rate for the control of surface condensation, unless of course a design safety margin has been built into the data used. The examples given thus far have shown that increasing the air change rate has the effect of reducing the relative humidity and hence condensation risk. It would be reasonable to assume that further increase in the air change rate would give progressive reductions in risk. This may not always be the case. The output of the heating system within a building will have a maximum value. This means that it will not be possible to maintain the design air temperature regardless of the size of the air change rate. This will have an adverse influence on the indoor relative humidity, as demonstrated in the following calculation, which is based on the procedure given in BS5250[5]. It should be noted that BS5250 presents this calculation procedure as a means of assessing surface condensation risk for a whole house, rather than merely using it to demonstrate the effect of ventilation rate on the internal relative humidity.

The heat balance for the building is given by:

$$Q = \Sigma AU(T_i - T_o) + \frac{1}{3}(NV)(T_i - T_o), \tag{3.6}$$

where T_i is the internal temperature (°C); T_o, the external temperature (°C); ΣAU, the total fabric thermal transmittance for the building (W/K); and N, the air change rate (air changes per hour).

The first component is the total fabric heat loss, whilst the second term is the ventilation heat loss.

Equation (3.6) can be rearranged to give:

$$T_i = \frac{Q}{[\Sigma AU + \frac{1}{3}(NV)]} + T_o. \tag{3.7}$$

It will be recalled from Equation (3.1) that:

$$g_i = g_o + \left(\frac{M}{\rho VN}\right).$$

By using the values of T_i and g_i calculated from Equations (3.7) and (3.1), the relative humidity resulting from the use of a specified air change rate can be calculated.

Example 3.6

Calculate the variation in relative humidity with air change rate for the following conditions:

- Internal conditions: moisture production is 5 kg per day (note that the internal temperature will vary according to the air change rate); volume of the space is 150 m^3; heat input is 1800 W and ΣAU is 120 W/K (*Note*: this includes surface resistances).
- External conditions: temperature is 5°C and relative humidity is 90%.

Assume that the density of air, $\rho = 1.2\,\text{kg/m}^3$.

Carry out the calculations for air change rates between 0.125 and 5.0 air changes per hour, incrementing in steps of 0.5 air changes per hour after the 0.125 air change per hour calculation.

Solution:

Use Equations (3.7) and (3.1) to generate internal temperatures and air moisture contents for the prescribed range of air change rates. From these values, relative humidities can be determined by one of the usual techniques. A graph of relative humidity against air change rate is shown in Figure 3.3. It can be seen that initially as the air change rate increases, the relative humidity decreases and with it the risk of mould growth. This is as a consequence of the moisture content of the internal air being reduced, and is predicted by Equation (3.1).

However, as the air change rate increases, Equation (3.7) shows that the temperature within the space will drop. Eventually, the decrease in internal temperature will overcome the decrease in moisture content,

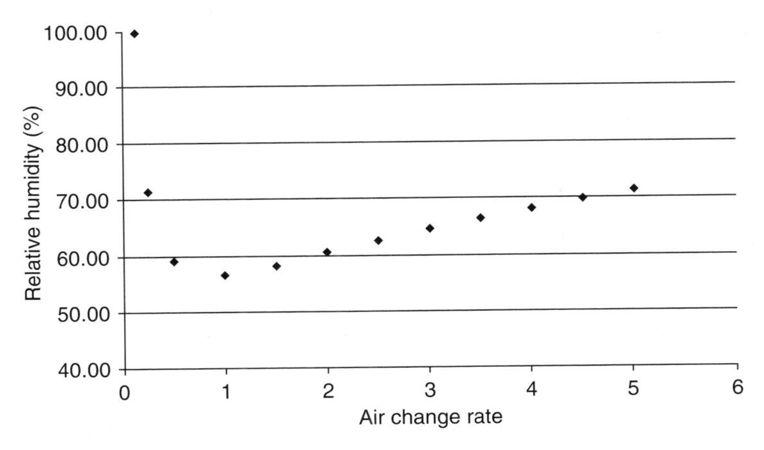

Figure 3.3 Variation of indoor relative humidity with air change rate

and as a result of this the relative humidity within the space will in fact rise. In this example, the turning point in the relative humidity. In this case, the maximum relative humidity occurs at about 1 air changes per hour. The position of the turning point will depend on a range of factors. However, BS5250 suggests that the range of optimum air rates will lie between 1.5 and 2.25 air changes per hour in the majority of cases.[5]

The results of the calculation show that progressive increases in air change rate will actually increase the relative humidity beyond a certain point, and could therefore make mould growth problems worse. This would be in addition to the reductions in the internal temperatures to the detriment of the comfort and the well-being of the occupants. This finding serves to underline the point that ventilation is not the only means of controlling condensation and mould growth, and that the provision of an adequate amount of heat is of great importance. Increase in the ventilation rate within a poorly heated dwelling will have no effect other than to make the unfortunate occupants colder and any mould growth problem worse. This is a very difficult point to grasp, but a vital one nonetheless.

As a note of caution, it is not always the case that the minimum ventilation rate calculated using Equations (3.6) and (3.7) will be appropriate. It may well be that none of the calculated solutions yield a relative humidity of less than 70%, and therefore mould growth will occur regardless of the ventilation rate. In such circumstances, consideration must be given to changing one or more of the other key variables in the determination of condensation of mould growth risk. The internal temperature may be raised; that is this may not be desirable from the comfort perspective. Although in many cases of existing problems, particularly within social housing, there is some scope for increasing internal temperatures, if they are low as a consequence of inadequate heating. The heat losses through the building fabric may be reduced; again this is quite feasible if the dwelling in question is a candidate for refurbishment. Finally, the discharge of moisture into the dwelling could be reduced, for example by not using unflued gas-heating appliances (portable liquefied propane gas heaters are particularly bad in this respect, not to mention a danger to occupant health and safety because of the emission of combustion products) or by venting tumble dryers directly to the outside.

3.1.5 *Interstitial condensation and the effect of air change rate*

The subject of interstitial condensation is not really within the scope of this book. However, in view of its potential for causing serious structural problems, it will be briefly discussed here, and the influence of ventilation on interstitial condensation risk is illustrated.

Interstitial condensation may occur within the structure of a building itself. It arises as a result of the structural temperature falling below the dew-point temperature at one or more points. This is illustrated in Figure 3.1. Interstitial condensation is most likely to take place during the winter, when internal and external differences in temperature and vapour pressure are at their most pronounced. Probably the most well-established method for predicting interstitial condensation risk is based on the Glaser method, a typical example of which is to be found within BS5250 (Appendix D).[6] When confronted with a range of values for a given material, care must be taken when selecting data for calculations, otherwise the risk of interstitial condensation may either be overstated or undetected.[7] This is more likely to be a problem with vapour resistivity data, since in practice it is the most difficult task to obtain repeatable results for vapour permeability measurements. The new version of BS5250 is rather critical of the limitations of the Glaser method for a number of reasons and seems to suggest that it might be better to make use of a dynamic simulation that combines heat flux, air flow and mass transfer (BS EN ISO 13788).[8] These are all well and good providing that the person with the design to assess not only has access to a dynamic simulation package but also understands how to use it. In the opinion of the author, the Glaser method is a perfectly adequate tool for "will it or won't it?" assessments, providing that care is taken with the selection of vapour permeability data.

The practical consequences of interstitial condensation taking place are dependent on the structure that is affected. Serious consequences are most likely to be observed in structures which include layers that are susceptible to water damage. A classic example is at the interface between insulation and plywood sheathing within a timber-framed wall construction, as shown in Figure 3.4. When condensation occurs, the water will soak the insulation, plywood and other timber. Under sustained exposure to high-moisture contents, mould growth will set in, with all the attendant risks for structural integrity. In theory, the vapour barrier in the structure should eliminate the risk of interstitial condensation. The reality is that controls on the building process within the UK are such that many vapour barriers are not put in correctly. They may not be correctly sealed at joints between adjacent sheets or at detailing points such as doors, windows and light switches. In the worst cases, the membrane may have been perforated during the construction process. This was particularly true during the building boom of the later 1970s and early 1980s, when timber-framed construction enjoyed an upsurge in popularity. The problems discovered were so substantial that the use of timber-framed construction came to an almost total halt. Public confidence in timber-framed construction is still very low, and mortgage lenders are most reluctant to become involved in the purchase of any dwelling so constructed. This is, in the opinion of the author, a matter of some regret, since soundly constructed timber-framed dwellings,

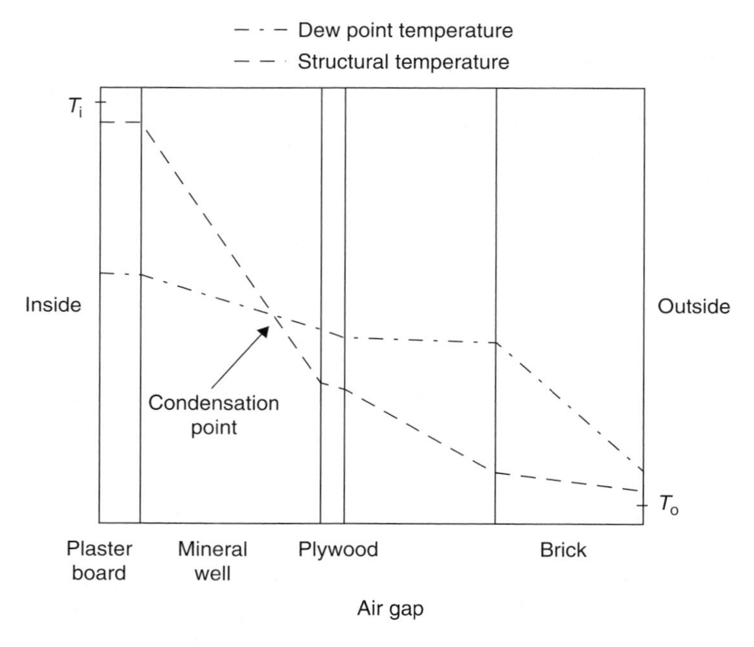

Figure 3.4 Interstitial condensation in a timber-framed wall

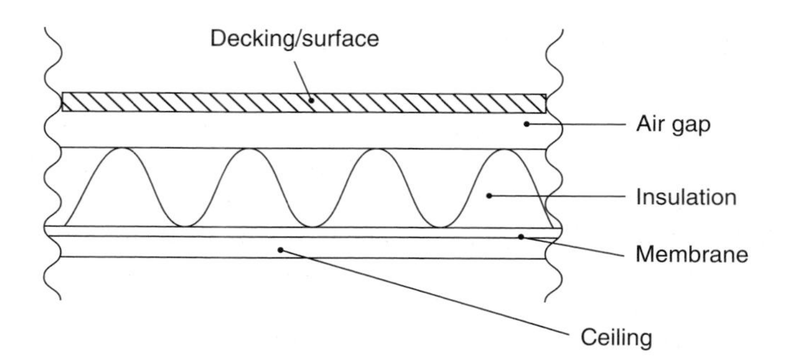

Figure 3.5 Typical cold deck flat roof cross section

for example following the methodology described in advised by the Timber Research and Development Association (TRADA),[9] offer great attractions in terms of speed of building, flexibility, energy efficiency and environmental impact of the construction materials used.

The other well-known structural element in which interstitial condensation is regularly reported, admittedly amongst a number of other failure mechanisms, is the so-called cold deck flat roof, as shown in Figure 3.5. These roofs have the most unhappy track record with respect to structural integrity, and it is extremely difficult to effect long-lasting remedial action other than replacement with another more robust roofing construction.

Happily the use of this structure has all but ceased within the heated zones of dwellings. It is now most likely to be encountered as part of a garage roof or else as the roof of an extension. With any luck the use of the cold deck flat roof will eventually be prohibited within the Building Regulations. For more information about the construction of flat roofs and associated technical problems, the BRE publication by Harrison[10] is recommended.

The issue of the influence of internal ventilation rate on interstitial condensation risk should be briefly considered. Without any doubt, it is possible to show that it is a simple matter to bring down internal water vapour pressure by means of increased ventilation, possibly to a level at which a Glaser calculation might indicate that there is no risk of interstitial condensation. However, it must be remembered that the Glaser procedure assumes steady-state conditions. In practice, this is not the case, and the risk is that some interstitial condensation might still take place. For this reason, the correct approach is to assume that the risk of interstitial condensation cannot be controlled merely by using a higher air change rate to reduce the internal vapour pressure within a space. Instead, the risk should be eliminated by means of careful attention to the design of structures and to the quality of construction detailing, particularly with regard to the installation of vapour impermeable membranes.

3.1.6 *Transient contaminant concentrations*

Analysis of changes in contaminant concentration with time is, for the most part, beyond the needs of most readers. Such calculations will involve the solving of differential equations, and are not suitable for a pen and paper/calculator approach. Dynamic simulations will make use of such calculation techniques, and the reader wishing to look at the transient behaviour is spared the effort of making.

A simple procedure, without recourse to mathematical proof, for the determination of contaminant concentrations after a given period of time is given in BS5295.[11] It can be shown that the variation of contaminant concentration in a space, given that the contaminant is being introduced into the space at a constant rate, is given by:

$$C(t) = \frac{(Qc_e + q)}{(Q + q)} + \left(1 - \exp\left(-\left(\frac{Q + q}{V}\right)t\right)\right), \tag{3.8}$$

where $C(t)$ is the concentration of the contaminant in the space at time t s from the moment at which the flow of the contaminant into the space begins; q, the flow rate of contaminant into the space (l/s); Q the volume air-flow rate of outside air into the space (l/s); c_e the concentration of contaminant in the outside air and Q/V is, of course, the ventilation rate in air changes per unit time.

As time progresses from the commencement of contaminant introduction, the term in the right-hand bracket will approach unity, and the value of $C(t)$ will consequently tend towards an equilibrium value C_{eq}, which is given by:

$$C_{eq} = \frac{Qc_e + q}{(Q + q)}. \tag{3.9}$$

The most important implication of this relationship is that the equilibrium concentration of pollutant is dependent only on the volume flow rate of outside air into the space and is thus independent of the space volume. What is influenced by the volume of the space is the time taken to reach equilibrium, which is dependent on the space volume; the larger the space, the longer the time for equilibrium. Equation (3.9) can be rearranged to give:

$$Q = \left(\frac{1 - C_{eq}}{C_{eq} - c_e}\right). \tag{3.10}$$

In this form, this equation offers an alternative approach to that given by Equation (3.2), which is presented in terms of the mass-flow rate approach.

In cases where the incoming air is free of the contaminant in question, then $c_e = 0$. This means that Equation (3.8) can be written as:

$$C(t) = \left(\frac{1}{(1 + Q/q)}\right)\left(1 - \exp\left(-\left(1 + \frac{Q}{q}\right)\frac{qt}{V}\right)\right). \tag{3.11}$$

Using Equation (3.11) in conjunction with systematic variation of the ratio Q/q gives a set of graphs of the form shown in Figure 3.6.

If there is no continuous input of contaminant into the space, then $q = 0$ and no observed change in $C(t)$ would be expected. However, if there was an initial concentration of contaminant, c_o within the space, then the rate of decay of concentration of the contaminant is given by:

$$C(t) = C(t = 0) \exp\left(\frac{Qt}{V}\right). \tag{3.12}$$

Equations (3.8)–(3.12) will be revisited during the course of the discussion of tracer-gas techniques given in Chapter 5.

In reality it is often not the case that a constant rate of contaminant emission will be observed. The contaminant may only enter the space for a limited length of time. In such a situation, the volume of the room

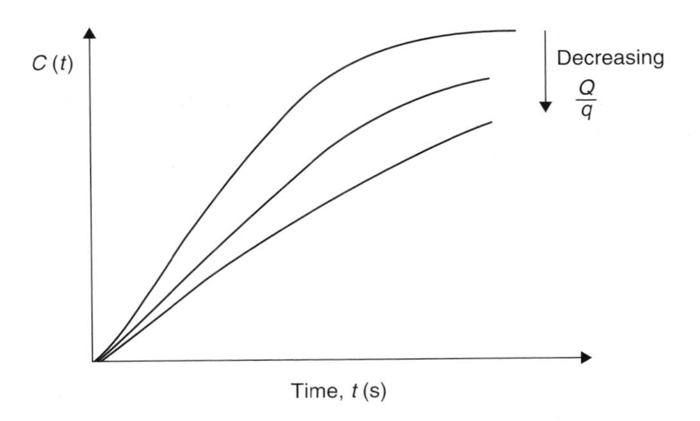

Figure 3.6 Influence of Q/q on the shape of the $C(t)/t$ curve

becomes an important factor in deciding whether the contaminant concentration reached will be an issue. For example, it might be important to know whether a permitted maximum permissible concentration was going to be exceeded in a particular case.

If the duration of the emission of contaminant is t_i, and if the space in question were to be ventilated at the rate Q as calculated by Equation (3.10), then the maximum concentration reached would fall short of the equilibrium value. The usual practice is to set the maximum permissible (or desired) concentration to the equilibrium concentration C_{eq}. In theory this means that in the case of intermittent contaminant input, the minimum ventilation rate can be recalculated as:

$$\frac{Q^F}{Q} = F\left(\frac{Qt_i}{V}\right), \tag{3.13}$$

where Q^F is the reduced minimum ventilation rate. The form of Equation (3.13) suggests that in cases where (Qt/V) is less than 1, no ventilation is required. In practice, it is most unlikely that such a situation would arise, as it will almost certainly be the case that, no matter how short the period t_i is, contaminant will be emitted later on.

Equations (3.8)–(3.13) can be used in the cases of a variety of indoor air pollutants, including water vapour.

Example 3.7

A room of $120\,m^3$ volume is heated by means of a flueless liquefied petroleum gas (LPG) appliance. The power output of the appliance is 2 kW. The ventilation rate within the room is 20 l/s. Normally, no more than three people would occupy the room when the heater is in operation.

Calculate the following:

(i) The CO_2 concentration within the space when 5 min have elapsed after the LPG appliance has been switched on and the occupants have entered the room.

(ii) The ventilation rate that would be needed in the room in order to maintain the equilibrium concentration of CO_2 at 0.5% v/v.

It should be assumed that occupancy by three people commences at the same time that the heating appliance is switched on. Furthermore, it should also be assumed that for an LPG appliance, the CO_2 production rate is 0.033 l/s/kW output, and that for each occupant, the rate of CO_2 production due to respiration is 0.004 l/s. The concentration of CO_2 in the outside air is 0.03% v/v.

Solution:

Firstly, the total rate of contaminant introduction into the space must be determined.

The total CO_2 production rate due to the LPG appliance, $CO_{2(LPG)}$ is given by:

$$CO_{2(LPG)} = 2 \times 0.0033 = 0.0066\,\text{l/s}.$$

The total CO_2 production rate due to occupancy, $CO_{2(OCC)}$ is given by:

$$CO_{2(OCC)} = 3 \times 0.004 = 0.012\,\text{l/s}.$$

Hence the total CO_2 production rate, $CO_{2(TOT)}$ is given by:

$$CO_{2(TOT)} = CO_{2(LPG)} + CO_{2(OCC)} = 0.0066 + 0.012 = 0.0186\,\text{l/s}.$$

(i) The concentration of CO_2 within the room after 5 min (300 s) can be found from Equation (3.8):

$$\begin{aligned} C(t = 600) &= \frac{(20 \times 0.0003 + 0.0186)}{(20 + 0.0186)} \\ &\quad \times \left(1 - \exp\left(-\frac{20 + 0.0186}{120}\right) \times 300\right) \\ &= 0.001221 = 1221\,\text{ppm} = 0.1221\% \end{aligned}$$

(ii) To find the ventilation rate necessary, Equation (3.10) should be used. In this case,

$$\begin{aligned} Q = 0.0186 \left[\frac{(1 - 0.005)}{(0.005 - 0.003)}\right] &= 9.2535\,\text{l/s} \\ &= 33\,\text{m}^3/\text{h} \\ &= 0.667 \text{ air changes per hour.} \end{aligned}$$

3.1.7 *Calculation of required ventilation rates for air quality*

In Chapter 2, the work of Fanger with respect to the assessment of indoor air quality was discussed. The approach used by Fanger is controversial, and certainly does not meet with the universal acceptance of all indoor air quality experts. Nonetheless, consideration of Fanger's methodology enables calculations of ventilation rates to be made with regard to what is required to provide a satisfactory level of perceived air quality, given knowledge about the sources and strengths of pollutants within the building in question.

There are two key units that have to be understood when applying the Fanger approach: these are the *olf* and the *decipol*. They are not wholly derived from the fundamental SI quantities; instead, they are partially based on the experimental observations of the perceptions of subjects of the strength of a standard odour. Those wishing to read more about the detail of the experiments should look at reference[12]. The *olf* is defined as the rate of odour generation form one standard person, working in an office or similar non-industrial workplace, sedentary and in thermal comfort with a hygienic standard equivalent to 0.7 baths per day! The *decipol* is defined as the perceived level of air pollution in a space with a pollution source of 1 olf ventilated by 10 l/s of unpolluted air.

The procedure is described in 14449 EN.[13] The actual technique is not simply to use the Fanger's methodology for the calculation. Rather, it is recommended that for any given situation, the ventilation rates required for health (control of risk associated with exposure to pollutant(s)) and for comfort (perceived air quality) should both be calculated and the higher of the two values should be used for design purposes.

The 14449 EN gives the following equation for calculating the required steady-state ventilation rate for meeting health requirements:

$$Q_h = \frac{M}{(C_i - C_o)\varepsilon_v}, \tag{3.14}$$

where Q_h is the ventilation rate for satisfying health requirements (l/s); M, the pollution load (generation rate) (μg/s); C_i, the permitted indoor air concentration of the pollutant in question (μg/l or parts per million); C_o, the pollutant concentration in the outside air (μg/l or parts per million) and ε_v, the ventilation efficiency (dimensionless parameter, as discussed in Chapter 2).

It will be seen that Equation (3.7) also has the same form as Equation (3.2), with different calculation units and the inclusion of the ventilation efficiency term.

Table 3.1 Outdoor levels of air quality*

		Air pollutants			
	Perceived air quality decipol	*Carbon dioxide mg/m³*	*Carbon monoxide mg/m³*	*Nitrogen dioxide μg/m³*	*Sulfur dioxide μg/m³*
At sea	0	680	0–0.2	2	1
In towns, good air quality	<0.1	700	1–2	5–20	5–20
In towns, poor air quality	>0.5	700–800	4–6	50–80	50–100

* The values for the perceived air quality are typical daily average values. The values for the four air pollutants are annual average concentrations.
European collaborative action. Indoor air quality & its impact on man. Environment and quality of life; Report No. 11, Guidelines for Ventilation Requirements in Buildings. Commission of the European Communities, Directorate General for Science, Research and Development, Joint Research Centre, Environment Institute, EUR 1449 EN, 1992.

The procedure for determining the required ventilation rate for comfort is more involved. The first decision involves the selection of a desired level of air quality that corresponds to a figure of 10%, 20% and 30% predicted dissatisfied occupants. Following this step, the perceived air quality of the outside air must be estimated using Table 3.1.

The third stage of the procedure requires the calculation of the sensory pollution load within the occupied space. Figure 3.7 shows values of pollutant load based on the level of activity and smoking behaviour. To calculate the total pollutant load per occupant, an occupancy density per square metre of floor area must be specified. Suggested values are given in Figure 3.8. The pollution load due to pollution sources that do not emanate from the occupants, such as furnishings and carpets, may be determined using the data given in Figure 3.9. Finally, it is necessary to estimate the ventilation efficiency for the system in use within the space. Typical values are given in Figure 3.10. These values may also be used in Equation (3.15). The ventilation rate for comfort purposes is then calculated from the following equation:

$$Q_C = \frac{10 \times M}{(CC_i - CC_o)\varepsilon_v}, \qquad (3.15)$$

where Q_c is the ventilation rate for satisfying comfort requirements (l/s); M, the sensory pollution load (generation rate) (olfs); CC_i, the perceived level of indoor air quality (decipols); CC_o, the perceived level of outdoor air quality (decipols) and ε_v, the ventilation efficiency (dimensionless parameter).

	Sensory pollution load (olf/occupant)	Carbon dioxide (l/h · occupant)	Carbon monoxide (l/h · occupant)	Water vapour (g/h · occupant)
Sedentary, 1–1.2 met				
0% smokers	1	19		50
20% smokers	2	19	11×10^{-3}	50
40% smokers	3	19	21×10^{-3}	50
100% smokers	6	19	53×10^{-3}	50
Physical exercise				
Low level, 3 met	4	50		200
Medium level, 6 met	10	100		430
High level (athletes), 10 met	20	170		750
Children				
Kindergarten, 3–6 years, 2.7 met	1.2	18		90
School, 14–16 years, 1–1.2 met	1.3	19		50

Figure 3.7 Pollutant loads and activity levels (*Source:* Reference[13])

	(Occupants/m^2 floor)
Offices	0.07
Conference rooms	0.5
Assembly halls, theatres, auditoria	1.5
Schools (class rooms)	0.5
Kindergartens	0.5
Dwellings	0.05

Figure 3.8 Occupancy densities (*Source:* Reference[13])

	Sensory pollution load (olf/m^2 floor)		Chemical pollution load TVOC (μg/s · m^2 floor)	
	Mean	Range	Mean	Range
Existing buildings				
Offices	0.3	0.02–0.95	–	
Schools (class rooms)	0.3	0.12–0.54	–	
Kindergartens	0.4	0.20–0.74	–	
Assembly halls	0.5	0.13–1.32	–	
Dwellings			0.2	0.1–0.3
Low-polluting buildings (target values)		0.05–0.1	–	

Figure 3.9 Pollutant load from non-occupant sources; TVOC: total volatile organic compounds. (*Source:* Reference[13])

Ventilation effectiveness in the breathing zone of spaces ventilated in different ways

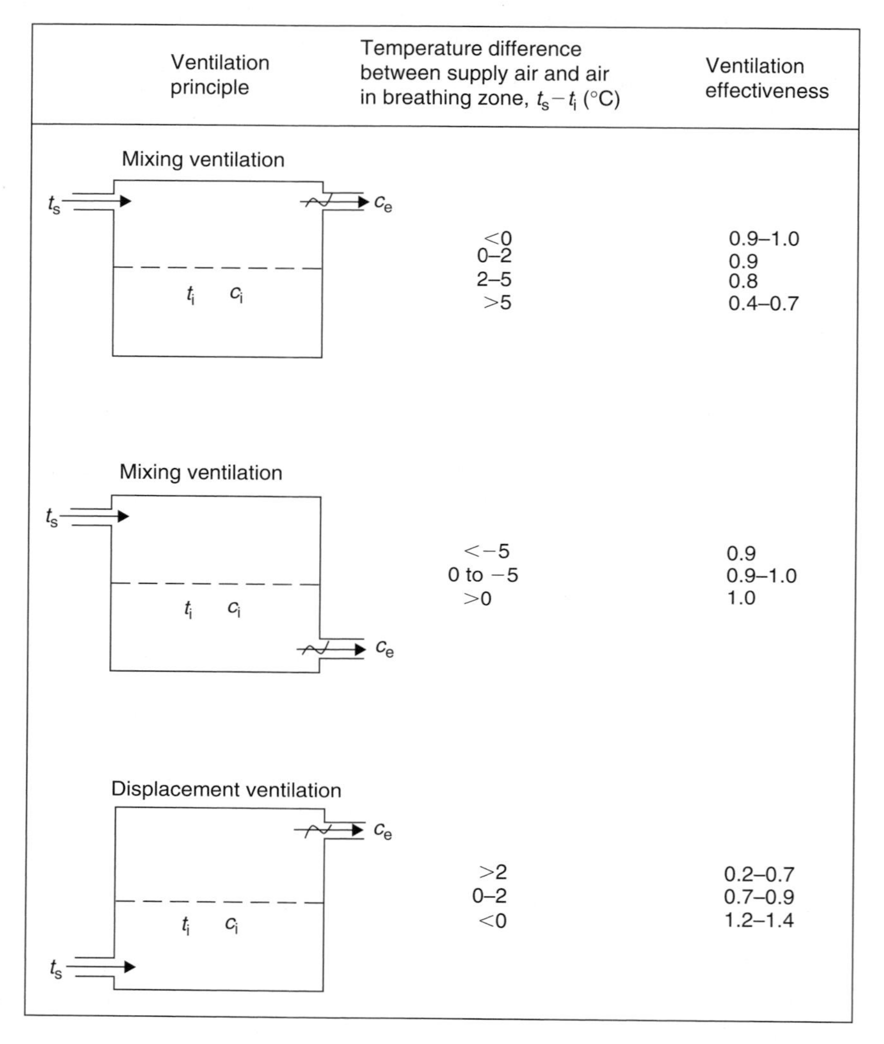

Ventilation principle	Temperature difference between supply air and air in breathing zone, $t_s - t_i$ (°C)	Ventilation effectiveness
Mixing ventilation	<0	0.9–1.0
	0–2	0.9
	2–5	0.8
	>5	0.4–0.7
Mixing ventilation	<−5	0.9
	0 to −5	0.9–1.0
	>0	1.0
Displacement ventilation	>2	0.2–0.7
	0–2	0.7–0.9
	<0	1.2–1.4

Figure 3.10 Estimation of ventilation efficiency (*Source:* Reference[13])

Example 3.8

An existing dwelling is located within a town where the outside air quality is good, and the perceived air quality is 0.1 decipols. Using the criteria, it is decided to try and achieve an indoor air quality of category B, corresponding to 20% of occupants dissatisfied and 1.4 decipols. None of the occupants of the dwelling are smokers, corresponding to 1 olf per occupant. The density of occupation (from Figure 3.8) is 0.05 occupants per square metre of floor area. In the absence of any data for dwellings, a value of 0.156 olf/m^2 of floor area is selected. A simple mixing ventilation strategy is being used. The ventilation efficiency is estimated as being 0.8. Calculate the ventilation rate required on the basis of that required to give the appropriate perception of air quality.

Solution:

First of all, the total olf load within the dwelling must be calculated. The contributions to the overall olf load are as follows:

- From the occupants: $1 \times 0.05 = 0.05$ olf/m^2.
- From the building itself: 0.05 olf/m^2.

Hence the total sensory pollutant load is $0.05 + 0.05$ olfs $= 0.1$ olf/m^2 of floor area.

From Equation (3.15), the required ventilation rate for occupant comfort corresponding to 20% of occupants dissatisfied is given by:

$$Q_c = \frac{10 \times 0.1}{1.4 - 0.1} \times \frac{1}{0.8} = 0.96\,\text{l/s/m}^2 \text{ of floor area}.$$

For a house of 300 m^3 volume and 50 m^2 total plan area, this would correspond to a ventilation rate of $0.96 \times 50 \times 3600 = 172.8$ m^3/h or 0.576 air changes per hour.

If the occupants were smokers, then the total sensory pollutant load would be increased to 0.2 olf/m^2, and the required ventilation rate would be increased to 1.92 l/s/m^2 (255.5 m^3/h or 1.15 air changes per hour).

The calculation technique is relatively easy to apply. However, there are several issues to take into account when assessing the usefulness of the results obtained. Much of the procedure relies on the data tables produced as a result of the work of Fanger. The controversy over this work should in itself serve as a note of caution. With respect to the data presented, it should be noted that no estimates of sensory pollutant load for dwellings are presented in EUR 14449 EN. A further complication to the domestic situation is the use of an air quality assessment system based on the percentage of dissatisfied occupants.

In a commercial building, where several hundred people might be in the workplace at any given time, an error of plus or minus five occupants would probably not be regarded as significant, given the semi-subjective nature of the assessment scheme. However, in a dwelling, an error of plus or minus one person could well be equal or greater to the percentage of dissatisfied occupants specified in the air quality classification system. For this reason alone, the use of the olf-based calculation system must be regarded as the highly suspect system within the context of domestic ventilation.

3.2 The prediction of ventilation rates

3.2.1 *Wind-induced pressure differentials*

There are two main driving forces which induce the pressure differentials needed for natural ventilation to occur. The first of these is the pressure difference due to the incident wind as it strikes the building and flows around it. The transfer of kinetic energy from the airstream to the building cause pressure differentials, and these pressure differentials are ultimately responsible for a component of the overall flow of air through the building. As the transfer is of kinetic energy, it can be said in general that at a surface on which wind is impinging, the pressure is a function of the square of the wind velocity.

Given the array of modern computing tools available for building design and simulation, it might be supposed that wind-induced pressure differentials might be easy to predict. Unfortunately this is presently not the case. When the wind hits a building, it flows around it. Only a small part of the building envelope actually receives the wind directly. Such areas are in a state of positive pressure relative to the inside of the building, as shown in Figure 3.11. The actual size and distribution of positive-pressure areas are influenced by roof pitch, as shown in Figure 3.12. For roof pitches greater than 30°, part of the leading edge of the roof will also be under positive pressure. The velocity of the air changes as it flows over the rest of the surface of the building. These areas are in a state of negative pressures relative to the inside of the building. The net result is the setting up of a regime in which there is the potential for air to flow into the building

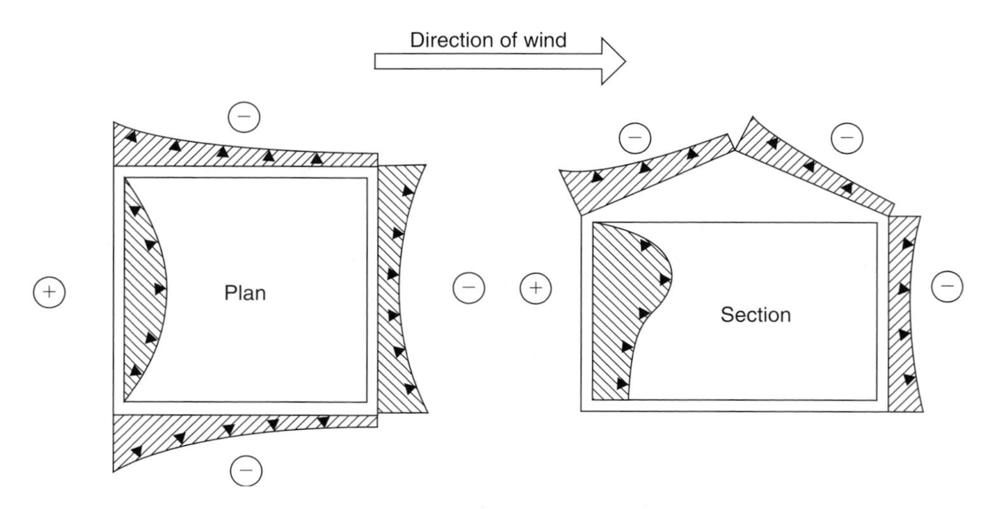

Figure 3.11 Air-flow pattern around a building as a result of incident wind

via the areas of positive pressure, and to leave the building via the areas of negative pressure. The word potential should be stressed. The actual magnitude of the resulting air infiltration is dependent on a number of other factors, and these will be discussed later.

The pressure regime around the building is influenced by the nature of the terrain in its immediate vicinity. If sited in an urban area, then the building would be afforded shelter from the impact of the wind by the adjacent buildings. The greater the number and size of the surrounding buildings, the greater the degree of shelter afforded. In general, this means that wind pressures around a dwelling in an urban area will be smaller than for a similar building in an exposed location. There are some exceptions to this generalisation. For example, in the case of tall buildings in close proximity to each other, the production of high-velocity wind vortices is a fairly common phenomenon. In rural areas, features such as trees and hedges may provide a limited amount of shelter. If the building is sited in a dip or fold, then some shelter will be afforded. Conversely, if the building is on a hill, then the building is more exposed, and the resultant wind pressures will be much higher.

3.2.2 *Measurement and calculation of wind velocity*

Wind pressure is proportional to the square of the wind velocity. Any error in the wind data chosen for an air infiltration calculation can, therefore, have a large effect on the result obtained. The issue is as follows: On what basis should wind data be selected for calculation purposes?

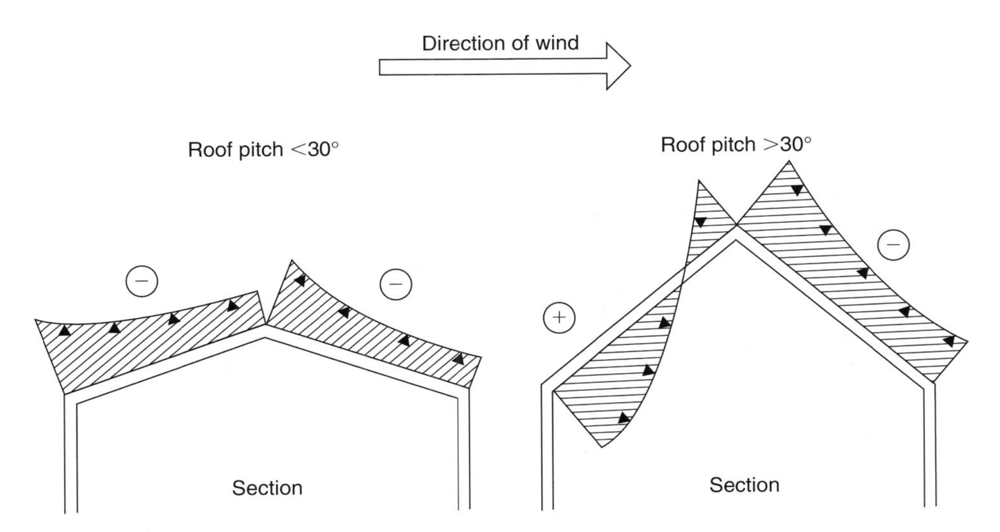

Figure 3.12 Air-flow pattern around a building as a result of incident wind – roof pitch greater than 30°

It has been seen that the wind pressure is proportional to the square of the wind velocity. Any error in the wind data chosen for an air infiltration calculation can, therefore, have a large effect on the result obtained. The basis for the selection of data for calculation purposes is an important matter.

The choice of how to obtain wind data for prediction purposes lies between one of the two routes: The first, and probably on first impression, the easiest is to use reference data for the site concerned. Wind data is available from several sources. The presentation of the data depends on the purpose to which the data is to be put. One such form of presentation is the isopleth chart, which shows contours of mean wind velocity for a prescribed time interval. Similar presentation can be used for the representation of maximum gust velocities within a given time scale. Such data for a 50-year time period is used in the calculation of uplift on pitched roofs BS5334.[14]

The processing of wind data for design purposes is currently the very topical research issue. It is widely recognised that there are dangers associated with the practice of merely choosing the wind data from the nearest weather station to the site of interest. The Chartered Institution of Building Services Engineers (CIBSE) amongst others have recently supported the research of Levermore *et al.* e.g. refer to reference[15] with the objective of significantly improving the wind data provided as part of the overall weather data within the CIBSE Guide J.[16] Indeed, a new version of the appropriate section of Guide J has recently been published. The intention is to both *improve the format* of the data in the guide, and also to *expand the number of sites* for which the data is available. Now the format of the guide has changed much rather than CIBSE providing Guide J as a self-contained resource, the actual weather data in a format suitable for use in calculation procedures described in the guide is supplied on a series of CD-ROMs (purchased in addition to the actual Guide J) in a form ready for importing into a spreadsheet or simulation package.

The second method of obtaining wind data is to carry out measurements at the site of the building in question. This has several major disadvantages. Of course the whole exercise will have to be carried out in real time; therefore, the collection of a statistically sound data set would take rather a long time to accumulate. If the building has not yet been built, then the lead into construction provides a valuable opportunity to collect data. If a short period of data monitoring is undertaken, then there is a risk that the data set may prove to be atypical. A period of 2 years of monitoring would probably be reasonable. In reality, such a period of time is unlikely to be available. The monitoring equipment needed is not particularly expensive in relation to many other types of scientific instrumentation, but nevertheless represents a capital outlay of several thousands of pounds. The equipment will usually be left unattended for most of the time, and therefore security issues arise.

In summary, whilst collection of data on site is a worthy ideal, in most cases it will not be practical to do so. Reliance will, therefore, be placed on tabulated data provided by such organisations as CIBSE.

Levermore *et al.*[15] deal with another interesting facet of wind data, namely the validity of velocity data collected at low wind velocities. It is quite natural to presume that the traditional cup anemometer gives reliable readings across its range. In fact, this is not the case. At low wind velocities, the inertia of the vanes and the rotating mechanism means that the measured velocity might be lower for short periods of time than the actual velocity. The way to avoid this is to use a more modern piece of equipment based on a thermistor array. This equipment has no moving parts and hence inertia effects are not an issue. Levermore *et al.* compare results from cup anemometers and thermistor arrays placed at the same locations. Notable deviations between the two types of sensors are reported at low wind velocities. In terms of natural ventilation, this equates to the underestimating ventilation rates. This could be very important when summer cooling of a building is the major concern.

3.2.3 *The interpretation of wind data*

In the majority of instances, it is not sufficient merely to collect wind data at a given site or else extract data from a reliable reference source. Careful consideration must be given to the vertical height at which the wind data was collected in comparison to the vertical height of the building that is under consideration. Wind velocity varies with vertical height. In general, wind velocity increases as vertical height increases. If data is being collected, or for that matter if information is being abstracted from sets of reference data, then the height at which the weather data was collected becomes an important issue. Any inaccuracy would cause large errors in the calculated wind pressures, with consequential errors in any calculated air change rates. The problem to be resolved is essentially that of extracting a value for wind velocity at roof height from a corresponding measurement or from a weather station on a mast of a known height.
The usual practice is to use a relationship of the form:

$$V_{\mathrm{roof}} = \mathrm{WRF} \times V_{\mathrm{mast}}, \tag{3.16}$$

where V_{roof} is the velocity at roof height; V_{mast}, the velocity at mast height and WRF is the *wind reduction factor* to generate the roof height value. WRF is in turn given by:

$$\mathrm{WRF} = K h_{\mathrm{roof}} a, \tag{3.17}$$

where K and a are dimensionless terrain-related constants, and h_{roof} is the height of the building roof.

Terrain type	*K*	*a*
Open flat countryside	0.68	0.17
Countryside with scattered wind breaks	0.52	0.20
Urban 0.35		0.25
City 0.21		0.33

Figure 3.13 *K* and *a* values (*Source: AIVC Guide*)

Values of K and a for a range of terrain conditions are given in Figure 3.13 (taken from reference[15]).

Of course, these figures are generalised. The responsibility for assessing the terrain type falls upon the user. Caution must be exercised. The form of Equation (3.17) reflects the influence of local terrain on wind velocities. It will be seen from Figure 3.13 that as the density of surrounding buildings increases, the recommended value of K decreases, and thus the overall WRF. When taking wind velocity measurements, the usual practice is to collect data at a height greater than the building being considered. For the majority of buildings, including houses, a mast height of 10 m will be adequate to achieve this. The 10-m-high weather mast has become something of a standard. When a taller building is involved, the mast would ideally be put at a taller height. The interpretation of weather data for tall buildings is not simple, and very great care must be taken when doing so. Fortunately, this will not be a problem for the majority of dwellings.

Example 3.9

Determine the roof height wind velocity for an 8.0-m-high building. The building is situated in the open countryside when the wind velocity measured at 10 m mast height is 3.0 m/s.

Solution:

Using Figure 3.13, for open countryside, $K = 0.68$ and $a = 0.17$. From Equation (3.17):

$$\text{WRF} = 0.68 \times (7.5)^{0.17} = 0.958.$$

Hence from Equation (3.16):

$$V_{\text{roof}} = 3.0 \times 0.958 = 2.874\ \text{m/s}.$$

3.2.4 *The determination of wind pressure distribution*

The first stage includes determining the direction and effective velocity of the wind. The next step in calculating the air infiltration rate within a building is to determine the pressure distribution arising from the impingement of the wind at a given velocity from a certain direction onto a building of a prescribed geometry. This is not an easy task. One possible means of obtaining this data set would be to carry out some measurements on an actual building. In practice, this is a very difficult thing to do. In most cases, ventilation will be considered at the design stage, which means that the building of interest will not yet be built. Unless an identical building exists to the one being considered, it would not be possible to carry out field measurements to yield the data needed. If the terrain around the building were to be different in nature to that around the proposed site, then any measurements taken would be of little value. In general, site measurements of this type are neither practical nor reliable.

The most widely used means of deriving data about pressure distributions are by the use of a scale model of the building in question inside a wind tunnel. The details of the use of wind tunnels for building pressure data generation are well beyond the scope of this book. Pressure monitoring points are located on the faces of the model, and measurements of pressure are taken for a range of wind velocities and wind directions. In some wind tunnels, the model is sited on a turntable, which allows for greater accuracy in the setting of the angle of the model with respect to the incident wind.

The most commonly used way of presenting wind tunnel data is as the sets of wind pressure coefficients. These are a very convenient way of accessing pressure distribution data. A typical set of data is shown in Figure 3.14. Tables for a range of typical building geometries are available in the AIVC Applications Guide.[17] Surface pressure coefficients are usually positive on the vertical surface of the building on which the wind is incident, and negative on the rest of the vertical surfaces. Pressure coefficients at roof surfaces depend on the roof pitch, but for pitches less than 30° are negative.

The pressure acting upon each surface of the building can be calculated from the equation:

$$P = c_{\mathrm{p}} \frac{1}{2} \rho v^2, \tag{3.18}$$

where P is the pressure (Pa); c_{p}, the wind pressure coefficient (dimensionless) and v, the wind velocity at the building height as calculated from Equation (3.16).

These are a number of assumptions which have to be made when using tabulated wind pressure coefficient data. It has to be assumed that there is no variation in the pressure coefficient regime as a result of turbulence effects at higher wind velocities. For example, in some cases in the vicinity of very tall buildings, this may not be a valid assumption. The tabulated data represents a mean value for the face of the building in question. For buildings of three storeys or less, use of a mean value is in almost all cases a valid assumption. However, in certain circumstances the use of the mean value may lead to problems; for example, in the case when ventilation

Low-rise buildings (up to three storeys)

Length to width ratio 1:1
Shielding condition Exposed
Wind speed reference level = building height

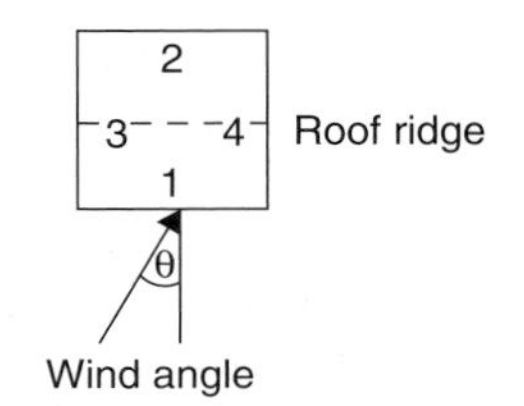

Wind angle

Location		0	45	90	135	180	225	270	315
Face 1		0.7	0.35	−0.5	−0.40	−0.2	−0.40	−0.5	0.35
Face 2		−0.2	−0.40	−0.5	0.35	0.7	0.35	−0.5	−0.40
Face 3		−0.5	0.35	0.7	0.35	−0.5	−0.40	−0.2	−0.40
Face 4		−0.5	−0.40	−0.2	−0.40	−0.5	0.35	0.7	0.35
Roof (<10° pitch)	Front	−0.8	−0.7	−0.6	−0.5	−0.4	−0.5	−0.6	−0.7
	Rear	−0.4	−0.5	−0.6	−0.7	−0.8	−0.7	−0.6	−0.5
Average		−0.6	−0.6	−0.6	−0.6	−0.6	−0.6	−0.6	−0.6
Roof (11–30° pitch)	Front	−0.4	−0.5	−0.6	−0.5	−0.4	−0.5	−0.6	−0.5
	Rear	−0.4	−0.5	−0.6	−0.5	−0.4	−0.5	−0.6	−0.5
Average		−0.4	−0.5	−0.6	−0.5	−0.4	−0.5	−0.6	−0.5
Roof (>30° pitch)	Front	0.3	−0.4	−0.6	−0.4	−0.5	−0.4	−0.6	−0.4
	Rear	−0.5	−0.4	−0.6	−0.4	0.3	−0.4	−0.6	−0.4
Average		−0.1	−0.4	−0.6	−0.4	−0.1	−0.4	−0.6	−0.4

Note: Approximate data only. No responsibility can be accepted for the use of data presented in this publication.

Figure 3.14 Pressure coefficient data (*Source: AIVC Guide*)

openings are located close to the ground. In such cases, the air flow through the openings may be underestimated (see Edwards and Hartless).[16] The degrees of allowance made for the sheltering effects of terrain are few in number. Great care must be taken in assessing the nature of surrounding terrain, given the semi-subjective classifications that are often used. If the air infiltration behaviour of the building under consideration is deemed to be critical to the success of the overall design, then it might be necessary to have the building modelled in a wind tunnel. This can be an expensive business. The number of wind tunnels available for measurements has diminished over recent years as research organisations have sort to reduce their operating costs. Fortunately, it is very rare to have to undertake such measurements for a dwelling. The process is more commonly used for tall buildings and groups of commercial buildings.

The need for the use of wind tunnels may eventually become a thing of the past. Computational fluid dynamics (CFD) may well prove to be a convenient and low-cost tool for the prediction of wind pressure coefficients. Such measurements should in principle provide a convenient means of integrating the determination of wind pressure coefficients with the rest of the air infiltration calculation process. However, the amount of computing power required is large, and the use of such software has until recently been completely beyond the resources of most architects and many engineers. Some very user-friendly packages are now available; one that immediately springs to mind is Star CFD. To run anything but a basic simulation, a very high-specification PC will be required. Commercial licences as opposed to the very favourable terms have been offered to academic institutions. At the time of writing, such a machine would be based on a twin xeon process or arrangement. For now, tabulated wind tunnel data will probably suffice for most users. In the mean time, users should not forget that CFD has other uses that go far beyond pressure coefficient prediction.

3.2.5 *Infiltration paths and ventilation openings*

In the typical dwelling, the building fabric provides an important route for air infiltration. The usual way of calculating air infiltration through solid elements of the building envelope is to use a power law representation of the variation in air flow with applied pressure, thus:

$$Q = AK\Delta P^{n}, \tag{3.19}$$

where Q is the air-flow rate (m^3/s); A, the area of solid element (m^2); and K and n, the coefficients specific to the construction of the solid element. Typical data is presented in Figure 3.15 (taken from reference[17], where much more data can be found).

Wall type	K	n
Unvented, clay brick with bare surfaces and granulated mineral wool insulation	0.024	0.81
Timber frame wall panel, with fibre board sheathing, rainscreen cladding and plasterboard lining	0.08	0.75
Expanded clay aggregate blocks with empty hollows	0.13	1.0

Figure 3.15 *K* and *n* values

The treatment of ventilation openings is more complex. In theory, every ventilation opening is different, with its pressure/air-flow characteristic a unique feature. In practice, trying to deal with the problem on this fundamentalist basis would render air infiltration modelling complex to the point of impracticality. Happily, assumptions can be made about the similarity of the behaviour of similar types of opening with little risk of causing significant calculation errors.

The most important issue to come to terms with is the actual air-flow behaviour associated with the particular geometry of opening. To understand this, it is necessary to consider two extremes of opening geometry. The first case is that of a large round opening, say of 100 mm diameter. This is typical of a through-the-wall ventilator. In such an opening, the air-flow regime is likely to be turbulent. The relationship between pressure and air flow can be described as:

$$Q = C_d A \left(\frac{2\Delta P}{\rho} \right)^{0.5}, \tag{3.20}$$

where Q is the volumetric air-flow rate (m^3/s); C_d, the discharge coefficient for the opening (dimensionless); A, the cross-sectional area of the opening (m^3); ΔP, the pressure differential across the opening (Pa) and ρ, the density of air (kg/m^3).

The relationship is known as the *common orifice flow equation.*

From Equation (3.20), it can be seen that for turbulent air flow, the relationship between pressure difference and air flow is non-linear in nature. The resulting increase in air flow is proportionately less as the pressure differential increases. This is a result of the increase in resistance within the turbulent air-flow regime. The presumption of a fully turbulent regime within the air-flow path may not be valid at low air flows. The critical parameter is the Reynolds number. The Reynolds number is given by:

$$\mathrm{Re} = \frac{DV\rho}{\mu}, \tag{3.21}$$

where D is the characteristic length; ρ, the density; V, the velocity and μ, the viscosity.

It is generally accepted that if the Reynolds number is above 3000, then the flow regime will be fully turbulent. The transitional region is between 2000 and 3000. Transitional conditions will exist in passive stack ducts at flows in the order of 15–25 m^3/h. The only effect of assuming a turbulent regime will be to slightly underestimate the air-flow rate.

The other extreme of pressure difference/air-flow behaviour of openings is the case of the air-flow path of extremely high aspect ratio. A good example of this would be the crackage around a doorframe. In such cases, the flow of air through the element would be laminar in nature, and could be represented by the *pipe flow equation*:

$$Q = \frac{\Delta P}{8\mu L}\pi r^4, \tag{3.22}$$

where μ, the dynamic viscosity and r, the radius of the opening. There might be some difficulty in setting a suitable value of r for some crack geometries.

It can be seen from Equation (3.19) that in this case the flow of air through the flow path varies in direct proportion to the applied pressure differential. In other words, the relationship is linear. Applied pressure differentials towards the higher end of the magnitudes experienced in the context of buildings will not result in a change to a turbulent regime.

Then these are the two extremes of pressure difference/air-flow path behaviour. In reality, the geometries and cross sections of many types of air-flow path vary at different points. This means that such openings will not exclusively exhibit one or the other of the two types of behaviour, but rather a mixture of the two. This leaves the air infiltration modeller with a serious dilemma. Given the number of air-flow paths within the typical building envelope, dealing with this extra complication would be difficult if not impossible. One particular difficulty is with the treatment of crackage type air-flow routes. At least large openings, such as mass-produced trickle ventilators and airbricks have fairly well-established pressure/air-flow characteristics, which vary little from one ventilator to the next.

The commonly accepted way of dealing with the issue is to assign to each air-flow element an air-flow characteristic of the form $Q = KA\Delta P^n$ as per Equation (3.20). In the case of manufactured ventilation openings, pressure/flow data will usually be available from the manufacturer, whilst in the case of crackage paths there are several sources of data, for example the AIVC Applications Guide.[17] Data of this type can be produced for the building envelope as a whole by means of fan pressurisation testing. For a detailed description of the fan pressurisation technique, refer to Chapter 4. This approach has been shown to give good correlation with the actual air-flow behaviour, although it has been noted by Etheridge[19] that in some cases a quadratic fit to the pressure-flow curve will give a better correlation at low-pressure differences. Given that all the other assumptions are made, omitting the use of quadratic law fits at low pressures will have little effect on the accuracy of air infiltration calculations.

3.2.6 *Buoyancy-driven air flows: the stack effect*

Wind-driven pressure differences are not the sole cause of air infiltration. Density differences between internal and external air masses can also produce pressure differentials. This section demonstrates how these pressure differentials arise.

The mechanism is not as straightforward as that for wind-induced pressure differentials. When the temperature of an air mass is uniform, the pressure of a mass of air at a height h (metres) above a given datum level h_0 (usually taken as either ground level or floor level depending on the circumstances) is given by:

$$p_\mathrm{i} = p_0 - \rho g h, \tag{3.23}$$

where ρ is the density of air (kg/m^3); g, the gravitational constant (m^2/s) and p_0, the pressure at the datum level (Pa).

It can be seen from Equation (3.23) that a linear pressure gradient will exist, with the pressure decreasing with increasing height above datum level. This is shown in Figure 3.16. The pressure gradient can be expressed as:

$$\frac{\mathrm{d}p}{\mathrm{d}z} = -\rho g. \tag{3.24}$$

Taking into account the gas laws, the gradient can also be expressed as:

$$\frac{\mathrm{d}p}{\mathrm{d}z} = -p_0 g \frac{273}{T}. \tag{3.25}$$

Thus it can be seen that the pressure gradient is inversely proportional to the absolute temperature of the mass of air under consideration.

It will almost certainly be the case that the mean air temperature within a dwelling will be different to that of the ambient air outside. The case of most interest is that the inside temperature is greater than that of the ambient air. This will also be the most common circumstance in the majority of dwellings for the heating season. In this case, since the inside air is warmer, its density will be less than that of the outside.

Consider the building shown in Figure 3.16. It has two openings in its external wall. These are at heights h_1 and h_2 above datum level (in this case, ground level). The two masses of air can be treated on the basis of

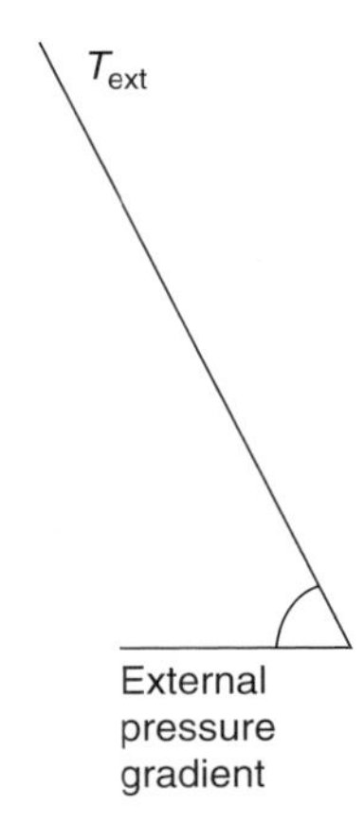

Figure 3.16 Vertical pressure gradients: building section (*Source: AIVC Guide*)

being two vertical columns separated by a vertical surface. The stack-induced pressure at height h_2 with respect to height h_1 is given by:

$$P_{stack} = -\rho g \times 273(h_2 - h_1)\left(\frac{1}{T_{ext}} - \frac{1}{T_{int}}\right), \tag{3.26}$$

where P_{stack} is the stack pressure (Pa); T_{ext}, the absolute external temperature (K); and T_{int}, the absolute internal pressure (K).

Typical variations in internal and external pressures for a simple building are shown in Figure 3.17. It can be seen that due to the different densities of the air masses, the pressure gradients are different. At lower heights, the external pressure is greater than the internal pressure. At greater heights, the internal pressure is greater than the outside pressure. This means that there is a potential for inside air to leave the building at high level, and for this air to be replaced by outside air to enter the building at low level.

This is the so-called *stack effect*, and it is very important in terms of the natural ventilation of buildings.

It will be noted that the two height/pressure lines intersect. At this point, the inside and outside pressures are equal. This point is referred to as the *neutral plane*. There is neither inflow nor outflow of air at this height. The position is of the neutral plane and is of some use in air infiltration modelling, as it affords a convenient means of allocating the total numbers and types of air-flow paths for inflow and outflow. However, it is not an essential parameter to determine.

The calculation of stack effect becomes more complex as the height of the building and internal partitioning increases. Tall buildings require particularly careful treatment. This subject is dealt with in some detail in the AIVC Applications Guide.[17]

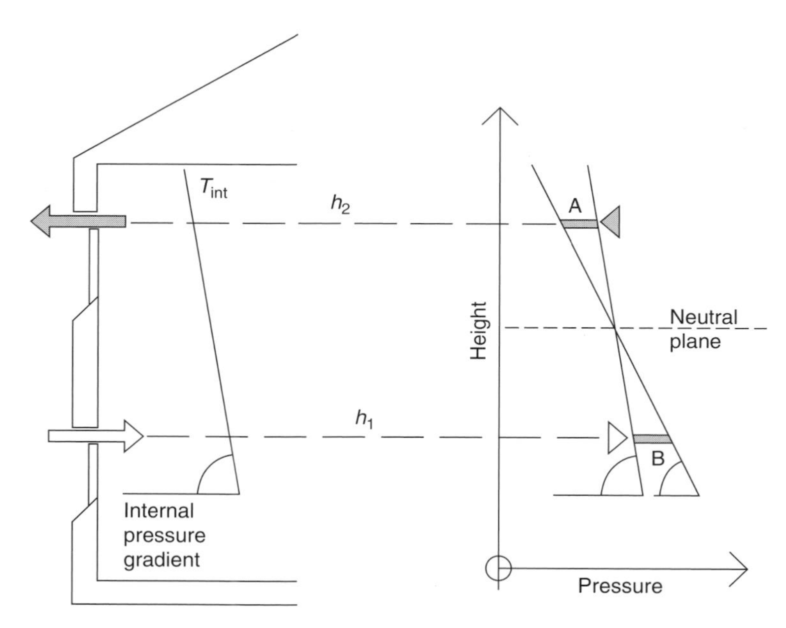

Figure 3.17 Internal and external pressure gradients: neutral plane and inflow/outflow

3.2.7 Combining wind- and stack-induced pressures

Pressure data is easy to handle when only one of the two pressure-generating mechanisms is in operation. In practice, stack and wind pressures will both usually be a component of the total pressure acting on the building. As might be expected, stack pressure dominates at lower wind velocities, whilst at higher wind velocities the wind pressure component prevails. The range of pressures resulting from stack effect will be smaller than that resulting wind effects. This is due to the fact that wind-induced pressure is a function of the square of the incident wind velocity.

The issue of the determination of air infiltration arising from the combination of stack and wind pressures is complex. It must be understood that the total air infiltration rate will not be the air infiltration rate calculated using the two pressure values separately. However, the total *pressure* at any ventilation opening is the sum of the stack pressure and the wind pressure. Combining the two pressures provides a convenient means of entering pressure data into simulation models.

Simplifying assumptions have to be made with care. BS5925[20] recommends that for the calculation methods contained within it, both the stack and wind components should be calculated, and the larger of the two used for the combined case. Such a treatment is too simplistic for more complex treatments of the prediction of air infiltration rates. In such procedures, the accepted means of dealing with the presence of stack and wind pressure components is to combine them in quadrature, in other words in the form:

$$Q_{tot} = (\Delta P^2_{stack} + \Delta P^2_{wind})^{1/2}. \tag{3.27}$$

This treatment is simple yet fairly reliable, and certainly more accurate than the BS5925 method.

3.2.8 *Procedure for calculating steady-state air infiltration rates: the influence of mechanical ventilation systems*

In some cases, it will be desirable to take into account the effect of mechanical ventilation on the overall air change rate within a dwelling. If the system is of the balanced supply and extract type, then there will be no overall interference with the air infiltration due to stack- and wind-induced effects. In the actual building, this assumption will only prove to be true if the system is correctly balanced when it is commissioned. In contrast, extract-only systems, for example kitchen and bathroom extract fans, will affect the air infiltration regime. The effect of extract ventilation can be taken into account very simply by consideration of the exponential air-flow equation as given in Equation (3.28). The pressure imbalance due to extraction is given by:

$$(\Delta P_{extract})^n = \frac{Q_{extract}}{K}, \tag{3.28}$$

where $(\Delta P_{extract})$, the pressure imbalance due to extraction; and K and n, the coefficients as derived from a fan pressurisation test.

In practice, this issue will be dealt with by most air infiltration modelling programmes, and will not be of any concern to the model user. Equation (3.28) can be used as a check on whether the extract fan will be able to deliver the required extract air-flow rate. This will not be possible if the fan cannot deliver the required flow of air at the calculated pressure imbalance.

3.2.9 *Single-zone calculation methods*

The use of single-zone methods is widespread. Their use is valid when the building is small, of limited height (typically three storeys or less) and where the user is content with the idea that all internal doors are open, which means that air flows from room to room are not likely to be throttled. The single-zone model can be either steady-state or transient. The two types of method have been discussed previously. Essentially, the air-flow routes in and out of the building are treated as a network, and an overall infiltration rate is calculated by determining the air flows in and out of each air-flow path. An overall mass-flow balance should exist. In other words, if there are n air-flow paths for a given building:

$$\sum_{i=1}^{n} M_i = 0, \tag{3.29}$$

where M_i is the mass flow of air through the ith flow path, which is in turn given by:

$$M_i = \rho_i Q_i, \tag{3.30}$$

where ρ_i is the density of the air flowing the ith element (kg/m^3) and Q_i the volumetric air flow through the ith element (m^3/h).

For further details on the determination of air infiltration rates, the AIVC Applications Guide[17] is recommended. This procedure is beyond the simple hand calculation methods, and a computer programme being preferred, but is probably appropriate for solution using a powerful modern spreadsheet. If an attempt was made to generate a mass-flow balance for a particular set of modelling parameters, then it is almost impossible to generate a balance on the basis of the parameters calculated in the first instance. Most models make use of iteration processes, where the calculated values are used as a first estimate of a solution, and the air flows are incrementally adjusted until a converging solution is obtained. Again, such techniques may nowadays be dealt with by means of the built-in analysis functions within the commercial spreadsheets.

3.2.10 *Simplified methods*

At the time when there was greatest research interest in the development of air infiltration prediction techniques, the late 1970s to the late 1980s, desktop computing was nowhere near as powerful as it is today. Due to this, there was a lot of interest in the development of prediction methods which contained major simplifications compared to the network calculation methods described above, yet did not sacrifice all the accuracy of these methods. The single biggest simplification that can be made is to eliminate the need for the calculation of air flows through individual openings.

The most well-known method within the UK is that based on the work of Warren and Webb[21] performed at the BRE in the late 1970s. At this stage, given the limitations on computer power prevailing, there was a particularly keen interest in estimating air infiltration rates without recourse to full-scale prediction techniques. The aim of the work of Warren and Webb was to allow estimates to be produced on a semi-empirical basis; in other words, from knowledge of a minimum number of physical parameters for a dwelling.

Central to the technique of Warren and Webb is the result of a fan pressurisation test. From this result, the air infiltration rate under the so-called "ambient" conditions can be estimated from the relationship:

$$Q = Q_T \left(\frac{\rho_0 V_r^2}{\Delta P_T} \right)^n F_v \ (A_r \phi), \tag{3.31}$$

where Q is the air-flow rate under ambient conditions (m^3/s); Q_T, the air-flow rate at a chosen reference pressure (m^3/s); ρ_0, the density of air (kg/m^3); V_r, the wind velocity at roof ridge height (m/s); ΔP_T, the internal and external pressure difference (Pa); F_v, the infiltration rate function; A_r, the Archimedes number and ϕ, the surface pressure function.

If wind-induced pressure is only considered, then Equation (3.31) can be expressed as:

$$Q_w = Q_T \left(\frac{\rho_0 V_r^2}{\Delta P_T} \right)^n F_w(\phi), \tag{3.32}$$

where F_w is the wind infiltration function.

For the case of stack-induced pressure only, Equation (3.31) becomes:

$$Q_s = Q_T \left(\frac{\Delta T \rho G h}{T_i \Delta P_T} \right)^n F_B, \tag{3.33}$$

where F_B is the stack infiltration function; ΔT, the internal and external temperature difference (K); G, the gravitational constant (m/s^2) and h, the height of building (m).

More details of the calculation of the infiltration functions F_B, F_w and F_v are given in Liddament.[17] Briefly, F_B is affected by building shape and the distribution of leakage paths, F_w is also influenced by the surface pressure coefficients for the building and F_v is influenced by wind velocity and temperature.

The Warren and Webb technique was validated by Liddament and Allen. It was found to give answers which were within ±25% of the measured values. This may not seem to imply a particularly good degree of accuracy, but within the context of ventilation modelling it is actually quite respectable.

3.2.11 *Calculating the air infiltration rate*

The procedure for calculating the air infiltration rate within a dwelling from basic physical parameters comprises a set of steps based on the previous sections of this chapter. These can be summarised as follows.

Preliminary steps

- *Gather basic building data*: plan area, height, type of construction, details of number type and distribution of ventilation openings.
- *Construct the data set for the calculation*: if the test data is not available, then values of pressure-flow characteristics will have to be selected.
- *Choose the environmental parameters for the calculation*: internal and external temperatures, wind velocity at reference height and wind direction.
- *Select the terrain type for the calculation.*
- *Choose the set of wind pressure*: coefficients for the building.

Calculation steps

- Use Equation (3.16) to determine the building height wind velocity.
- For each face, determine the incident wind pressure using Equation (3.18).
- Determine the stack pressure within the building.
- Calculate the total pressure across each face of the building.
- For each side of the building and its roof, determine the air infiltration through the fabric and its direction.
- For each ventilation opening, determine the air-flow rate and its direction.
- From the air flows determined, calculate the total air infiltration rate.

In theory, the calculation procedure should yield a simple mass-flow balance, in which the mass of air entering the building should be equal to that leaving the building; in other words the relationship:

$$\Sigma Q = 0. \tag{3.34}$$

However, mainly as a result of all the assumptions made during the creation of the data set for the calculation, it is almost certain that a balance will not be achieved, and as a result there will be a discrepancy between the inflow and outflow mass-flow rates. Such a situation would be contrary to the laws of physics, and this leaves the modeller with a major problem. More accurate data would reduce the discrepancy but could never eliminate it. What is much more useful is a robust means of using such data as is available to produce a credible figure. Recourse must be made to a mathematical treatment of the basic inflow and outflow data generated. On the basis that the output produced lies somewhere near the actual state of affairs, the most common course of action is to make use of an iteration process. The values of elemental mass-flow rates produced by the initial calculation are used as "first estimates" in the iteration process. On every cycle of the iteration, the values are adjusted slightly, either upwards or downwards, so that at every iterative step, the value of the sum of the air mass flows approaches a value which if not actually zero is so close as to make no difference. This is known as *convergence*. Unless there is something fundamentally wrong with the initial data set used, then achieving convergence should not present any problem.

Whilst the use of iteration is necessary to produce a mathematically consistent set of flow data, there is one big snag, namely that the use of calculation procedures of even this level of sophistication is turned into an exercise which is beyond what can be handled by paper and calculator by the majority of interested parties. Spreadsheets can be used to carry out iterative processes, but great skill in the use of the spreadsheet is necessary in order to achieve this. The end result is that those who wish to achieve quick estimates of air change rates will usually seek to do this by other more simple means, such as those described in Section 3.2.10 above. Those needing the output and accuracy of the more involved calculation procedure will almost inevitably resort to the use of computerised prediction tools. These are discussed in more detail in the next section.

3.2.12 Computer simulation packages suitable for domestic applications

AIDA

Air Infiltration Development Algorithm (AIDA) is a simple programme produced by Martin Liddament, formerly of the AIVC and now of VEETECH. This is not actually a full-simulation package, but rather a short piece of BASIC computer code. It will produce answers in its original form, but is not really user friendly, as knowledge of the sequence of the input variables is needed. However, with very limited knowledge of BASIC programming, it can be transformed into an easy-to-use prediction tool that can be customised to the individual users' requirements.

BREVENT

BREVENT was developed by the BRE in 1986. A revised package specially designed for ease of use, was issued by the BRE in 1988. The software is designed to run on a fairly low specification PC under Microsoft Disk Operating System (MS-DOS). To the author's knowledge, there have been no recent updates of the programme into a more contemporary appearance, neither are there seemingly any plans to do so. Whilst it achieves its purpose very well, the whole appearance of the software therefore looks very dated in comparison to modern software based on Visual BASIC and Windows. The graphics within the package are particularly outmoded. Countering the issue of the age of the programme, it must be pointed out that BREVENT has two extremely useful features. Instead of individual data sets having to be run one by one, BREVENT has the capability to run sets of simulations over specified ranges of internal and external temperatures, wind velocities and directions. The output of the simulation can be diverted to spreadsheet readable files. This is a very useful feature to have, as it permits further analysis of large amounts of air infiltration data.

The programme has a simple system for data entry. The user has to do nothing more than make sure that appropriate data is entered via a set of data entry screens under the headings on a well-thought out menu tree. The programme sets sensible upper and lower values of most parameters, and warns the user if an erroneous value has been entered before the user leaves the data entry screen in question.

The programme can accommodate the use of extract fans and passive stack ducts, and also enables the user to simulate the effect of chimneys and open flues.

BREVENT has what on first sight a rather odd feature, namely the ability to calculate the air infiltration rate in a sub-floor cavity beneath the dwelling being modelled. The reason for this that BREVENT was adapted to provide a prediction capability for a major piece of Building Establishment research work concerned with the control of radon ingress into dwellings (Hartless *et al.*).[22]

It should be noted that the sub-floor calculation procedure has been successfully validated using site data derived from tracer-gas decay measurements (Edwards and Hartless).[23]

3.2.13 *BS5925 calculation methods*

BS5925[11] presents a technique for the calculation of natural ventilation rate in what the standard refers to as simple buildings. No precise definition of a "simple building" is offered. The method is essentially based on the methods presented above, but with a number of simplifications. These are achieved by the use of a number of data tables and figures given within BS5925 itself.

At the heart of the calculation method, there are two tables of equations for calculating natural ventilation air flows. One table gives equations for cross-flow ventilation regimes, whilst the second table gives similar equations for single-sided ventilation. These tables are reproduced as Figures 3.18 and 3.19. Of the two, Figure 3.18 is likely to be of more interest within the domestic context.

In order to be able to calculate natural ventilation air flows, several key data items must be determined. An important data item for calculations involving cross-flow ventilation is the equivalent leakage area of the ventilation openings in the building envelope. Presumably for ease of calculation, the figures accompanying Figure 3.18 only show four openings, two on each side of the building. There is no reason why more should not be used. The rule is that openings in parallel (i.e. in the same elevation of the building) may be added together arithmetically, whilst those in series (i.e. openings on opposite elevations of the building) may be combined as the sum of the reciprocals of their squares.

Natural ventilation of a simple building		
Conditions	Schematic representation	Formula
(a) Wind only	u_r, A_1, A_3, C_{p1}, C_{p2}, A_2, A_4	$Q_w = C_d A_w u_r (\Delta C_p)^{1/2}$ $\frac{1}{A_w^2} = \frac{1}{(A_1 + A_2)^2} + \frac{1}{(A_3 + A_4)^2}$
(b) Temperature difference only	θ_e, θ_1, H_1	$Q_b = C_d A_b \left[\frac{2\Delta\theta g H_1}{\bar{\theta}}\right]^{1/2}$ $\frac{1}{A_b^2} = \frac{1}{(A_1 + A_3)^2} + \frac{1}{(A_2 + A_4)^2}$
(c) Wind and temperature difference together	u_r, θ_e, C_{p1}, θ_1, C_{p2}	$Q = Q_b$ for $\frac{u_r}{\sqrt{\Delta\theta}} < 0.26 \left[\frac{A_b}{A_w}\right]^{1/2} \left[\frac{H_1}{\Delta C_p}\right]^{1/2}$ $Q = Q_w$ for $\frac{u_r}{\sqrt{\Delta\theta}} > 0.26 \left[\frac{A_b}{A_w}\right]^{1/2} \left[\frac{H_1}{\Delta C_p}\right]^{1/2}$
Note: It should be appreciated that, in practice, some openings exits unintentionally, e.g. junctions between building components, and that such openings will contribute to the ventilation actually achieved.		

Figure 3.18 BSI Table – cross-flow ventilation equations

The equation for the calculation of wind-induced ventilation rates makes use of a simplified treatment of surface pressure coefficient data. This is much simpler than the surface pressure coefficient data sets normally required to make calculations of air infiltration rates. In the case of wind-only cross-flow ventilation, as dealt with in Figure 3.18, use is made of what is termed as the applied differential mean pressure coefficient, or ΔC_p. BS5925 presents a table of pressure coefficient data for a range of building height-to-width ratios, and plan ratios of a rectangular nature. Data is given for incident wind angles of 0° and 90°. This table is reproduced in this chapter as Figure 3.20. ΔC_p is simply calculated as the arithmetic difference between the two surface pressure coefficients for the two building faces under consideration. For example, in the example shown in Figure 3.21, the value of ΔC_p is $+0.7 - (-0.25) = +1.0$. BS5925 makes the point that, since the calculated ventilation rate will be a function of the square root of ΔC_p, then for a range of applied differential mean pressure coefficients between 0.1 and 1.0, a 10-fold increase will only result in a corresponding range of ventilation rates that increases by a factor of 3.

Natural ventilation of spaces with openings on one wall only		
Conditions	Schematic representation	Formula
(a) Due to wind	Plan	$Q = 0.025Au_r$
(b) Due to temperature difference with two openings	A_1; θ_e; H_1; $\Delta\theta + \theta_n$; A_2	$Q = C_d A \left[\frac{\varepsilon\sqrt{2}}{(1+\varepsilon)(1+\varepsilon^2)^{1/2}}\right]\left[\frac{\Delta\theta g H_1}{\bar{\theta}}\right]^{1/2}$ $\varepsilon = \frac{A_1}{A_2}$; $A = A_1 + A_2$
(c) Due to temperature difference with one opening	H_2	$Q = C_d \frac{A}{3}\left[\frac{\Delta\theta g H_2}{\bar{\theta}}\right]^{1/2}$ If an opening light is present $Q = C_d \frac{A}{3} J(\phi)\left[\frac{\Delta\theta g H_2}{\bar{\theta}}\right]^{1/2}$ where $J(\phi)$ is given in figure 7

Figure 3.19 BSI Table – single-sided ventilation equations

Figure 3.22 also gives an equation for the calculation of ventilation rates due to temperature difference only. This calculation is quite straightforward, and makes use of only the internal and external temperature differences and the vertical spacing between openings.

BS5925 gives a complex verbal explanation of the factors influencing ventilation rates when both wind- and temperature-induced ventilation effects are combined. However, the chosen solution to the problem is fairly straightforward. Depending on the value of the ratio of incident wind velocity to the square root of internal and external temperature difference, the overall ventilation rate will be calculated either as that due to wind-induced ventilation, or as that due to temperature-induced ventilation rate alone.

3.2.14 *Multi-cell modelling*

Treating the house as a single cell is appropriate in many cases, especially those in which the need is to look at overall air change rates for purposes of

Mean surface pressure coefficients for vertical walls of rectangular buildings							
Building height ratio	Building plan ratio	Side elevation/plan	Wind angle α (degrees)	C_p for surface			
				A	*B*	*C*	*D*
$\frac{h}{w} \leq \frac{1}{2}$	$1 < \frac{l}{w} \leq \frac{3}{2}$	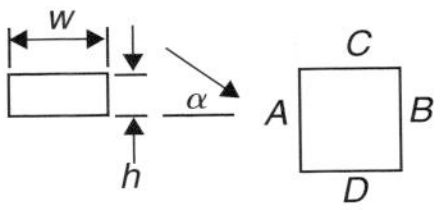	0	+0.7	−0.2	−0.5	−0.5
			90	−0.5	−0.5	+0.7	−0.2
	$\frac{3}{2} < \frac{l}{w} < 4$	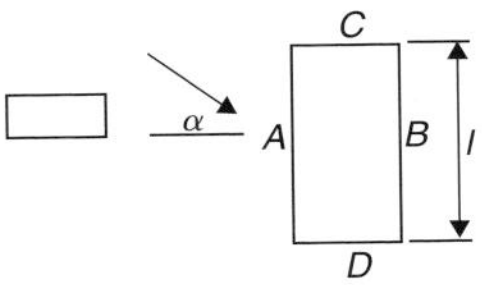	0	+0.7	−0.25	−0.6	−0.6
			90	−0.5	−0.5	+0.7	−0.1
$\frac{1}{2} < \frac{h}{w} \leq \frac{3}{2}$	$1 < \frac{l}{w} \leq \frac{3}{2}$	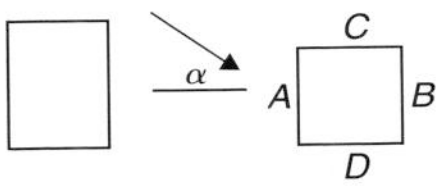	0	+0.7	−0.25	−0.6	−0.6
			90	−0.6	−0.6	+0.7	−0.25
	$\frac{3}{2} < \frac{l}{w} < 4$	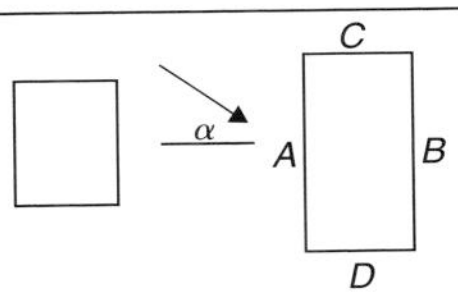	0	+0.7	−0.3	−0.7	−0.7
			90	−0.5	−0.5	+0.7	−0.1
$\frac{3}{2} < \frac{h}{w} < 6$	$1 < \frac{l}{w} \leq \frac{3}{2}$	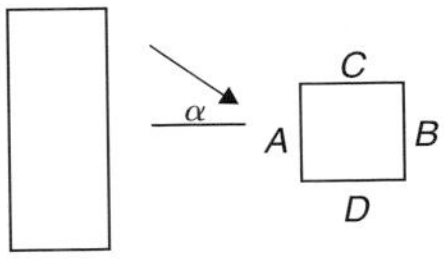	0	+0.8	−0.25	−0.8	−0.8
			90	−0.8	−0.8	+0.8	−0.25
	$\frac{3}{2} < \frac{l}{w} < 4$	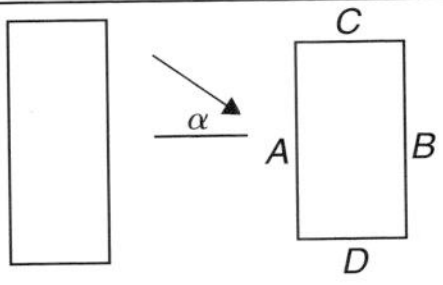	0	+0.7	−0.4	−0.7	−0.7
			90	−0.5	−0.5	+0.8	−0.1

Figure 3.20 Mean surface pressure coefficient data (Table 13 in BS5925)

controlling condensation in the medium to long term, and where there is no real interest in localised effects within individual rooms. In some cases, this may not be appropriate. In some energy studies, there is a need to estimate the flow of air from heated zones to unheated zones. In some condensation studies, the movement of moisture from zones of generation to

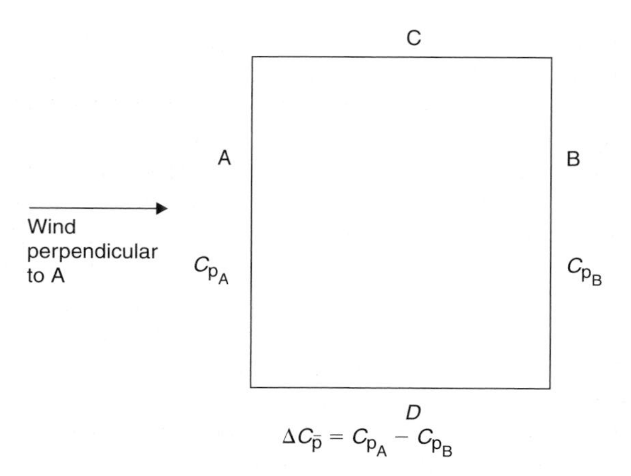

Figure 3.21 Principle of applied differential mean pressure coefficient. *Note*: in relation to C_{p_A}, C_{p_B} will he negative

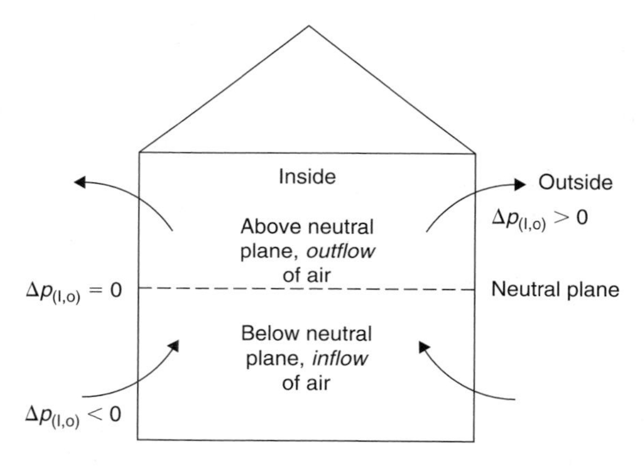

Figure 3.22 Internal and external pressure gradient: neutral plane and inflow/outflow

other areas can be of importance with respect to condensation risk, especially when the other zones are at lower temperatures.

As might be expected, the means of predicting air flows within multi-cell systems are much more complicated than single-volume systems. Instead of the relatively simple situation of a single volume with one air temperature, we are now confronted with a network of volumes, all with different internal temperatures and degrees of connectivity with adjacent cells. The solution required therefore changes from a simple mass-flow balance over the external envelope to a mass-flow balance between all the cells, with the mass balance over the envelope superimposed. If the whole system is thought of as a network, then the outside could indeed be thought of as a cell in its own right.

The mathematical procedures for deriving inter-cell air flows become more complex as the number of cells increases. Therefore, one of the first steps in setting up the modelling procedure would be to decide how many cells are required, being very careful not to specify any more cells than are absolutely necessary. Once this is done, the air-flow network which is between must be defined, with all the direct interconnections specified. Of course, it will not be the case that all cells will be connected, much less in two directions. For example, unless there are problems with cavity closers in a house of timber-framed construction, it is highly unlikely that there will be a direct air-flow path from downstairs to the roofspace in a two-storey property, and even less probable that a path will exist in the opposite direction. Once this has been done, all the characteristics of individual flow paths have to be specified. This will be a very involved process, and there is little scope for simplification.

Once the data set has been specified, a calculation procedure is followed which has some similarities to the single-cell case. A first calculation will be made of all inter-cell air flows. What will result is a set of data which will in all likelihood be considerably out of balance. The only way forward is an iterative process, with every risk that the solution may not converge for a specified data set.

Results from multi-cell models are much more sensitive to the input data than those derived from simple single-cell models. Given the extra date input required, this is not surprising. Much of the input data, such as the air leakage of partition wall and the crackage area around "closed" doors will be estimated. On balance, such models may be of value in resolving complex issues of pollutant movement.

References

1 H Stymne, M Sandberg, O Holmgren. Ventilation measurements in large premises. *Proc. Roomvent '92, Vol. 2. Aalborg, Denmark*, 1992.
2 BS5250 Psychrometric chart. Control of condensation in buildings, page 53, BSI 1989.
3 ibid pp. 58–61.
4 Thermal Insulation: avoiding risks. Building Research Establishment, 1992.
5 Air change/humidity information from BS5250. Control of condensation in buildings, Appendix D, pp. 66–76. BSI 1989.
6 BS5250 Interstitial condensation calculation (Appendix D).
7 KM Letherman. The choice of vapour resistivity values in calculations of condensation risk, Vol. 7, No. 1, 1986, pp. 49–52.
8 BS EN ISO 13788.
9 TRADA Timber framed housing guide.
10 HW Harrison. *Roofs and Roofing – Performance, Diagnosis, Maintenance, Repair and the Avoidance of Defects*, 1996; pp. 58–61, pp. 135–150. BR302. ISBN 1 86081 068 3.
11 BS5925 Code of Practice for ventilation principles and designing for natural ventilation BSI, 1991.

12 PO Fanger. Introduction of the old and the decipol units to quantify air pollution perceived by humans indoors and outdoors, *Energy and Buildings*, 1988; 12, 1–6.
13 EUR 14449 EN.
14 BS5334 Code of Practice on Slating and Tiling BSI, 1997.
15 G Levermore. Inaccuracies in standard meteorological low speed wind data. Proceedings of the ASHRAE Symposium, Cincinnati 2001.
16 Guide J: Weather, solar and illuminance data (CD-ROM), CIBSE, 2002.
17 MW Liddament. *Air Infiltration Calculation Techniques – An Applications Guide*, 1986.
18 RE Edwards, R Hartless. *AIVC Conference Paper*, 1993.
19 DW Etheridge. Air leakage characteristics of houses – a new approach. *Building Serv. Eng. Res. Technol.* 1984, **5**(1), 32–36.
20 BS5925 Natural Ventilation, 1992.
21 PR Warren, BC Webb. *AIVC Conference*, Windsor, 1980.
22 R Hartless, MK White. Air Infiltration and Ventilation Centre. "The Role of Ventilation". Proceedings of 15th AIVC Conference. Buxton, UK, 27–30 Septemer 1994, Vol. 2, pp. 687–696.
23 RE Edwards, R Hartless. *AIVC Conference Paper*, 1993.

4

Measurement Techniques

4.1 Introduction

No matter how high the reliability of modern air-infiltration prediction techniques are, there is still a need to be able to carry out *in situ* measurements. Problems such as inadequate internal temperature, excessive energy consumption and surface condensation will arise when they would not necessarily be expected. Prior to the monitoring of the effectiveness of an energy-conserving refurbishment programme on a group of dwellings, it is useful to have an indication for the ventilation heat losses taking place.

In this chapter, the useful techniques for the determination of important ventilation parameters are reviewed. Where they are likely to be of practical interest, they are described in detail. Finally, some aspects of *in situ* monitoring are also discussed.

4.2 Tracer-gas techniques

A brief mention of the equipment requirements for tracer-gas measurements is appropriate. In all site measurements, the identity of the tracer gas will be known. There is no need to use an analytical instrument which is actually able to identify the tracer gas, merely one which can measure its concentration would be enough. This rules out the need for the use of sophisticated laboratory techniques, such as mass spectrometry, which are not in any case suitable for use in the demanding fieldwork environment. The most popular types of equipments are probably infrared gas analysers and probable gas chromatographs.

The use of tracer-gas techniques goes back to over 50 years. The basic principles remain the same. A gas that can readily be detected is released into the volume of air under study, and its variations in concentration, either with time or position, are noted. The use of the technique was not limited to use in buildings; indeed, some tracer-gas tests are intended to yield information about large-scale air movements within the atmosphere over distances of hundreds of kilometres or more. Some of the methods used were given the gift of hindsight, rather dubious; one might speculate about the health and safety implications of using hydrogen or radioactive gases as tracers.

Exhaustive treatments regarding the mathematical theory of tracer-gas measurements have been presented in several other texts, and it is not proposed to repeat the exercise here. Readers, who wish to find out more about this particular aspect of tracer-gas measurements are advised to consult the technical publications of the *Air Infiltration and Ventilation Centre* (AIVC). Instead, this chapter will concern itself with actual methods, and will describe the mathematics only to the appropriate level of detail required.

In practice, there are some limitations on the type of tracer gas that can be used. Tracer gases should ideally have to be used only in small amounts. They should not be toxic or injurious to health by other means. This precludes amongst others, the use of radioactive tracer gases and nitrous oxide. The latter was a popular choice during the 1970s, but has since been shown to pose as a health risk to unborn children at quite low atmospheric concentrations. A tracer gas should be chemically stable once released. If it reacts with any other substance, then any measured concentrations will not be representative of the air change rate. It would ideally not be present in ambient air, although this need not prove to be a huge problem in reality, provided the background concentration is reasonably stable. The environmental effects of tracer gases must, nowadays, be taken into account. During the 1970s and 1980s, chlorofluorocarbons (CFCs) were the popular tracer gases. Nowadays, the capability of these gases to damage the earth's ozone layer is well understood, and the use of freons as coolants, even in small amounts needed for tracer-gas tests, is now regarded as unacceptable. Indeed, it is not now possible to buy new stocks.

Due to the perceived complexity of tracer-gas measurements, which in the author's opinion is not entirely justified, their use is not common at the present. Of the limited amount of activity that is taking place, the most popular tracer gases seem to be sulphur hexafluoride, the perfluorocarbon group of compounds and carbon dioxide (CO_2). The latter is of course a by-product of respiration and combustion, and this presents a risk of producing spurious data. On the other hand, with knowledge of levels of occupancy, metabolic CO_2 is a useful way of introducing tracer gas into a room.

There are three principle groups of tracer-gas techniques. Constant concentration techniques aim to sustain a target concentration of tracer gas within a space and to monitor its concentration. When tracer-gas concentration falls below the preset target value, injection of more tracer gas is automatically triggered. For a good description of the use of the technique refer to reference[1]. Simple air change rate determinations can be undertaken (single tracer gas) simply by monitoring the rate of tracer-gas input. Interzonal air movements can similarly be determined by the use of multiple tracer gases, although recirculation of gases becomes an issue. Modern equipment and computing power have made such techniques a viable proposition. Among the equipments available, the Bruel and Kjaer proprietary system is probably the most well known. Constant concentration equipment is extremely expensive and is only within the budgets of well-equipped research organisations and university departments. The major technical difficulty in the use of concentration techniques is stabilising the gas concentrations. In some situations where mixing is not very good, this may be very difficult to achieve. The use of mixing fans during tests may be necessary. There is some uncertainty regarding the effect of this on internal air-flow patterns within the test building, particularly where the building is naturally ventilated. Rapid fluctuations in incident wind velocity, which in turn cause fluctuations in the actual air change rate will make the attainment of steady-state conditions more difficult.

Constant emission techniques rely on the monitoring of the variations in tracer-gas concentration with time when tracer gas is trickled into space at a constant rate. Older methods resulted in very large concentrations of tracers, which meant that they were not suitable for occupied buildings. The most popular modern variation of the constant emission method is the passive sampling tube technique. This type of measurement was pioneered by Dietz *et al.*,[2] and was further developed and applied by a number of workers, most notably Perera *et al.*[3] This can be used to measure air change rates on a medium- to long-term basis. The tracers used are from the perfluorocarbon group. Gas is introduced into the test cell by diffusion through a rubber cap fitted to a glass or metal vial. The rate of diffusion is remarkably steady throughout the useful life of the tracer-gas source. The concentrations of tracer gas are very low, typically parts per million or less, and they can be used quite safely in occupied spaces. The tubes and tracer-gas sources are relatively inexpensive, but the gas chromatography equipment needed to analyse the contents of the sampling tubes is expensive, and is certainly not intended to endure the rigours of transport to, from and in some cases across the test sites. Sample tubes are usually returned to the laboratory for analysis. This leads to the risk of damage to, and consequent contamination of sample tubes, leading to unreliable results. Constant emission is not a very successful method of determining interzonal air movement, but may be used to give an overall long-term indication of air movement rates.

Decay techniques are probably the most frequently used type. A pulse of tracer gas is injected into the space and mixed in with the air, often by the use of oscillating desk fans, in order to produce a uniform concentration. The exponential decay of the tracer-gas concentration with time is monitored and the air change rate is found by making a best fit to the experimental data points of the form:

$$\log_e C_{(t)} = \log_e C_{(t=0)} + \exp(-nt), \tag{4.1}$$

where $C_{(t)}$ and $C_{(t=0)}$ are the tracer-gas concentrations at time t and $t = 0$, respectively, n is the air change rate.

Air change rate is usually expressed as air changes per hour, which is often abbreviated to ach for convenience. Variations of the technique have been used to measure air change rates within wall[4] and sub-floor cavities.[5] The decay technique has been adapted by several researchers. For example, Edwards and Irwin[6] have used a multiple tracer-gas technique for determining interzonal air flows for up to four interconnected cells. Such a capability is achieved at the cost of increasingly sophisticated monitoring equipment and mathematical analysis; indeed, it should be noted that there has been some debate about the most appropriate mathematical treatment of concentration data collected during multiple-zone tracer-gas measurements.[7]

4.3 Site monitoring

According to the order in which appropriate assessment of the significance of measured air change rates, it will usually be necessary to measure several supplementary parameters. These will usually include:

- wind velocity,
- wind direction,
- internal temperature,
- external temperature.

The most commonly used type of temperature sensor is probably the platinum resistance thermometer (PRT), suitably encapsulated for protection. The technology has not advanced very far in the past decade. The choice of wind velocity measurement probe may be important. For general-purpose use, the standard rotating-vane anemometer will be satisfactory. If there is interest in building performance at low-incident wind velocities, then the inertia of the rotating-vane mechanism represents a potential source of measurement error. In such circumstances, one of the modern (but rather more expensive) ultrasonic measurement systems should be used instead.

In studies involving condensation risk, relative humidity measurements will also be useful. Energy-oriented studies will also have their own specific parameter monitoring requirements.

The frequency of the measurements should be selected in accordance with the objectives of the tracer-gas study.

In cases where ducted ventilation systems are installed, it may be useful to monitor air flows within the ducts. There is a good choice of suitable equipment available for this purpose.

Monitoring, recording and analysing large amounts of data have become much easier over recent years, whilst the cost of the associated equipment has not increased significantly. In commercial buildings, it is often possible to download data directly from the buildings' BEMS (Building Energy Management System). Such a technology is very rare in dwellings. Even then, significant savings on travel to and from test buildings can be achieved by downloading data from monitoring equipment by means of telephone and/or Internet connections.

4.4 Air leakage testing

As has been previously mentioned, tracer-gas measurements are time consuming, expensive and usually require skilled personnel. In many cases, they are not necessary. An indication of the likely air infiltration through a building can be found, not by making measurements of the air change rate within a building, but rather by the air leakage characteristics of the building envelope. Two types of techniques may be used: they are the direct current (DC) fan pressurisation and alternating current (AC) pressurisation techniques.

4.5 DC fan pressurisation

This type of test is often referred to as simply *fan pressurisation*, for no other reason that the other type of test, AC pressurisation, is not frequently used for *in situ* measurements. The DC pressurisation test is often referred to as a *blower door test* in American or Canadian literature. In the UK, a popular source of guidance is that of Stephen and Webb[8] of the Building Research Establishment (BRE). At the time of writing (late 2003), it seems inevitable that a requirement for a fan pressurisation test of a new dwelling will soon be incorporated into the Part L of the Building Regulations for England and Wales.

The essentials of the measurement technique are shown in Figure 4.1. The basic equipment includes the following:

- *A mobile fan (preferably portable)*: This should be capable of delivering sufficient air up to an applied internal–external pressure differential of about 60 Pa. Stephen and Webb[8] state that 55 Pa would be sufficient. For an average UK property, a fan capacity of 4000 m^3/h should be adequate. Axial fans are usually favoured on the grounds of ease of handling, although centrifugal fans may be used; 240 V AC single phase is

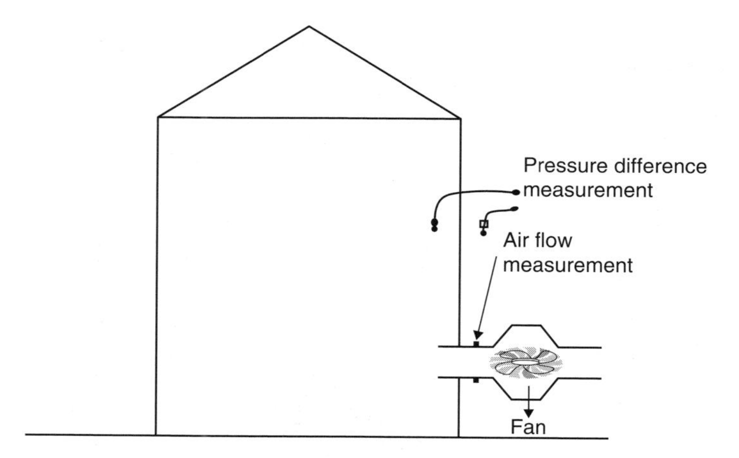

Figure 4.1 Fan-pressurisation testing schematic

the most convenient power source, be it from mains or generator. Some fans have direct-drive petrol motors. Fans may be obtained as a part of blower door test kit, as shown in Figure 4.2, or else be incorporated into a bespoke piece of apparatus. Readers should be assured that the blower door is by far, the most convenient for site work. However, bespoke pieces of apparatus can de designed so as to have a larger cubic capacity. The author has access to a piece of equipment, which uses the most powerful single-phase axial fan that can be purchased. Fan pressurisation equipment is available for tests on buildings larger than dwellings, but such an equipment is very large and difficult to transport. The ultimate piece of equipment is BREFAN, which can be used to pressurise medium-sized commercial buildings. The equipment should be capable of both positive and negative pressurisation.

- *An air-flow-measuring device*: Stephen and Webb specifically recommend against the use of fan characteristic data. Any air-flow-measurement device used should in their opinion have an accuracy of better than ±5%. If the chosen device does not conform to the appropriate British Standards,[9,10] then the device should be calibrated *when incorporated into the assembled test apparatus*, and then it should be calibrated by the use of standard orifice plates in accordance with the requirements of reference[11]. The unit shown in Figure 4.3 incorporates a Wilson-flow grid. This is in effect a multi-point averaging device which is factory calibrated so that values of air-flow rate can be obtained simply by measuring the average pressure drop across the flow grid.
- *A pressure-measuring device*: The traditional inclined manometer serves this purpose very well. However, there are alternatives available such as dial gauges and electronic manometers, which may be favoured on grounds of ease of transportation and setting up under

Figure 4.2 Blower door

site condition. The pressure is measured between the outside and inside of the building. Siting of the two pressure measurement points must be done with care. For instance, if the flow of air through the fan system impinges on either of the measurement points, then a false result will be obtained. In most cases, it will be convenient to connect the pressure-measuring device to the measurement points by means of flexible tubing. Care must be taken to ensure that the tube selected is not excessively flexible, or else it will have a tendency to compress and hence deform its cross section, as shown in Figure 4.4. This will also lead to false readings, which also tend to fluctuate erratically as the tube moves.

Figure 4.3 Fan pressurization equipment incorporating Wilson flow grid

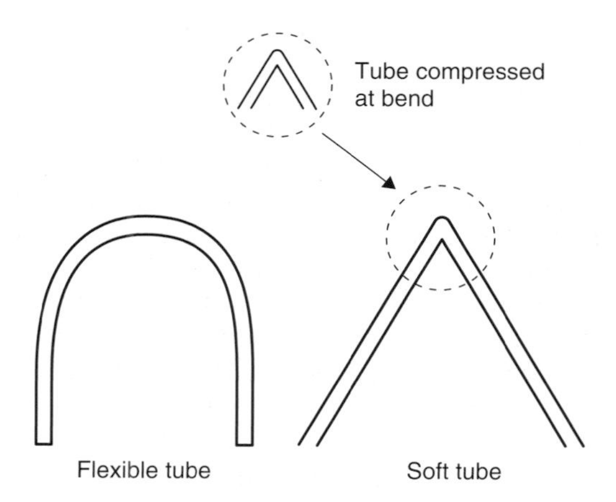

Figure 4.4 Compression of soft flexible tubing

- *Air-flow control system*: The simplest method is to incorporate a mechanical damper (preferably an iris damper) into the fan system. The use of such a device will have implications for air-flow distribution within the delivery system and may even necessitate the use of a flow-straightening element. Another possibility is to use a fan with variable blade pitch. Off-the-shelf blower door units usually make use of an electronic-speed controller for ease of handling and transportation. The unit shown in Figure 4.4 incorporates both an iris damper and an electronic-speed controller in order to give the unit a wider range of air-flow delivery rate and a more sensitive control.

- *A means of attaching the fan system to the building being tested*: The most convenient way of doing this is to use a piece of board which fits directly into an open door or window. The board is held in place with sealing tape and possible small nails. This method has several drawbacks. Not all doors or windows of the same nominal size prove to be exactly the same. This means that if more than one of the same type of building is being tested, then some degree of adaptation will be required using small pieces of board and/or adhesive tape. If the building has been decorated, great care has to be taken when using adhesive tape in order that serious damage is not inflicted. Off-the-shelf blower door units are usually mounted on a special mounting board system which has a suitable range of dimensional built-in adjustment. The edges of such boards are usually finished with a replaceable self-adhesive foam draught strip. This gives a very good seal when the door is correctly adjusted, and in addition minimises the risk of decorative damage.

- *Connecting ductwork*: In some situations, it may not be possible to make a direct connection to a door or a window. An example of this might be in a case where restricted space exists around a doorway. A way around this would be to use a piece of connecting ductwork. For convenience this would probably be of the spiral-wound flexible variety. For the typical fan used for these tests, the diameter of the duct will be of the order of 400 mm. Care needs to be taken when using flexible ductwork on site. If the duct is accidentally perforated, then false readings will be produced. The resulting errors will be small, but will nonetheless be errors. The duct must not be distorted if bends are necessary, otherwise large reductions in available air flow will occur as a result of the extra resistance. If the duct moves around during tests, then it will be difficult to get steady readings of pressure difference and air flow. The most important thing to bear in mind is that if the fan pressurisation system is intended for use with a variety of ductwork configurations, then its design must ensure that the calibration of the air-flow-measuring device is not invalidated.

- *Air temperature measurement device*: Volumetric air flows are influenced by air temperature. If a significant internal–external temperature difference exists, then correction for air-density difference will have to be made. This means that a temperature-measuring device should be available. An accuracy of ±0.5°C will be adequate.

Other pieces of equipment can be used but are probably in the "nice to have" category. For example, if electronic air-flow- and pressure-measuring instruments are being used, it is very easy to interface these to a laptop computer and record data directly.

Prior to tests, some preparatory work is required. If the houses are already occupied, liaison with the occupants is essential. They must understand as far as possible, what is going to happen. The author usually tries to give them the idea of a trip to the shops. This way, they will not be tempted to do frustrating things such as open and close doors and windows during tests. If the houses have solid fuel or coal fires/burners, the occupants must be warned not to have them burning. This would be potentially quite dangerous. It is always worth checking oneself, prior to the commencement of tests.

Care must be taken to seal flued appliances (switching off pilot lights on gas appliances would be prudent), chimneys and purpose-provided ventilators closed off prior to testing. With some ventilators, sealing out with low tack tape and/or plastic sheeting may be required. The loft hatch, unless it is held in place by a latch system, may start to rise from its normal position under positive pressure. It is often advisable to seal out loft hatches, or better still, held down with a weight. A brick fulfils this purpose admirably. Of course, it is important not to forget to remove the brick after the test, otherwise the next person who attempts to enter the loft might be injured. In new dwellings, particular care should be taken to ensure that the traps on sanitary appliances are filled with water, otherwise significant air leakage will take place through them.

The main purpose of the fan pressurisation test is to determine the airtightness of the building fabric. However, this is not to say that there is not some value in repeating tests with various openings unsealed. By comparison with the baseline case of all openings sealed, useful data can be obtained. It all depends on whether the house is being measured with the objective of generating useful research data, or else if the primary interest is to produce a figure, indicative of the fabric. In the latter case, it is likely that the tests are being carried out with an eye on the clock. In these circumstances, the strategy is to do the bare minimum of work on each test property, and hence needless test work will be frowned upon. When the schedule of test work entails sealing and unsealing of openings; great care must be taken to ensure that the work is done in a systematic manner. This will minimise risk of errors in the recording of test details and measurement results. The method recommended by Stephen and Webb[12] is to progressively seal building components and/or openings of interest when positive-pressure testing is taking place. When the building is depressurised, the components and/or openings should be unsealed in reverse order. This procedure is referred to as "reductive sealing".

Upon arrival on site, the strength of the wind in the vicinity should be assessed: if it appears that the wind strength is greater than force 3, Stephen and Webb[13] recommend that test work does not proceed, since it will be very difficult to obtain stable pressure/air-flow measurements due to wind-related pressure fluctuations. They also caution that if measurements are made under wind strengths between Beaufort 2 and 3, the accuracy of the results obtained will be adversely affected.

In the absence of a detailed schedule of test house characteristics, it is advisable to systematically collect data about the house in a manner that has similarities to the methodical approach used by a building surveyor. Dimensions of the property must be measured, and the volumes of internal cupboards determined so that the calculated volume of the house can be corrected accordingly. Some thought may have to be given to the actual house volume under test. For example, in the case of an attached garage, this would not normally be included in the test volume. Careful recording of house details is usually well rewarded. A sample of 10 properties on a new estate may nominally be of the same specification, but in practice, there will be variations in details, and in some cases, variability of the quality of workmanship in such areas as window sealing and draught stripping. This will result in a spread in the measured air-flow rates for the properties. It is very useful to be in a position to account for variations in results on the basis of observations of defects in the building fabric recorded at the time of the tests.

The rate of air flow into the building is measured at each pressure difference in the case of positive pressurisation of the building. The test procedure is then repeated, but this time the rate of air flow out of the building is measured for depressurisation. This is done because it will almost always be the case that the pressure/air-flow data results obtained for the pressurisation and depressurisation tests will be different. This is due to the fact that envelope components such as doors and windows distort by different amounts under depressurisation as compared to pressurisation and hence allow different amounts of air to pass as a result of the extra leakage areas produced, as shown in Figure 4.5. Under positive pressure, doors will tend to be pressed more firmly into their frames, thus reducing air leakage. The opposite would normally be expected for windows. A loft hatch will, of course, have a tendency to give an increased leakage area, unless they have been properly held down or sealed out as described above. Regardless of whether the test is being conducted under negative or positive pressure, care should be taken with some types of air inlets and

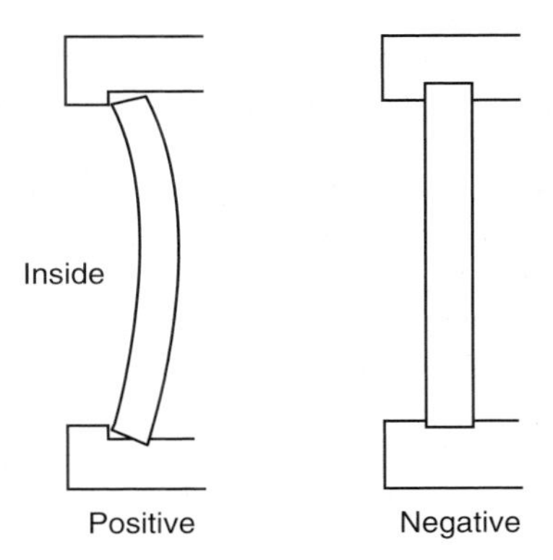

Figure 4.5 Window distortion under depressurisation

outlets as the sealing vanes/dampers on them, even if in the closed state, have a tendency to fly open beyond a certain applied pressure differential, thereby giving artificially elevated readings of air flow beyond this pressure threshold.

The significance of the positive-pressurisation test versus depressurisation test has been discussed at length in the scientific literature. At the moment, the accepted practice when processing test results (in the absence of any sound evidence to the contrary) is to effectively average the two sets of results provided that neither of them contains any anomalous readings, in keeping with the recommendation of Stephen and Webb.[14]

The treatment of results differs according to the purpose of the tests. If the internal and external temperature difference is greater than 2.5°C, then it is recommended that an air temperature correction factor is applied in order to account for density differences. Stephen and Webb[15] give two correction equations, depending on whether the test in question is pressurisation or depressurisation.

For pressurisation:

$$Q_{corr} = Q_m \frac{T_o + 273}{T_i + 273}. \tag{4.2}$$

For depressurisation:

$$Q_{corr} = Q_m \frac{T_i + 273}{T_o + 273}. \tag{4.3}$$

where Q_{corr} = corrected airflow (m^3/hr); Q_m = measured airflow (m^3/hr); T_i = Internal temperature (°C) and T_o = Outside temperature (°C).

In circumstances where data sets are being collected merely to provide a value of the standard measure of airtightness, a simple graph plotting is acceptable. The pressurisation and depressurisation test results are plotted as two curves on one graph, as shown in Figure 4.6, and the mean value of the air leakage rate at 50 Pa, the so-called Q_{50} value is measured from the graph. If the objective of the test is to produce useful data say, for example, ventilation simulation, then a more involved treatment of the raw data is required. For each set of test results, a best fit is made to the data points of the form:

$$\log_{10} Q = \log_{10} C + n \log|\Delta P|, \tag{4.4}$$

where Q is the air-flow rate (m^3/h), ΔP, the applied pressure difference (Pa) and C and n are the co-efficients.

Taking antilogarithms:

$$Q = C\Delta P^n. \tag{4.5}$$

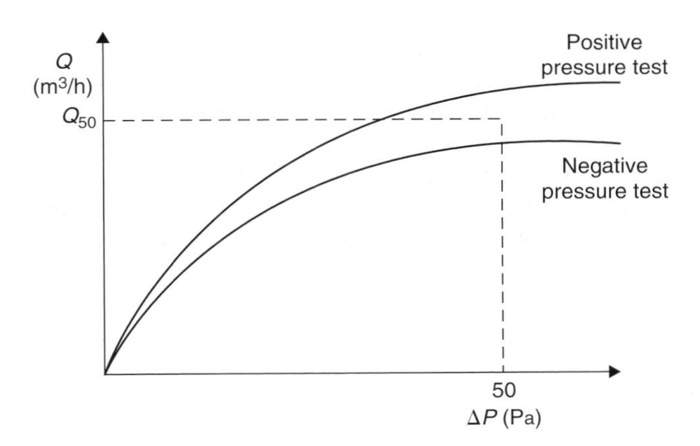

Figure 4.6 Determination of mean Q_{50} from graph

Given the great improvements in ease of processing of data afforded by such software tools as spreadsheets, the use of the curve-fitting approach has become far easier than it was even 5 years ago. It is the author's impression that most field measurements seem to make use of the curve-fitting approach.

The co-efficient C is an indication of the overall crackage area of the building in question, whilst n varies according to the nature of balance between pressure/air-flow regimes within air leakage routes. If $n = 1$, then this would imply that air flow was wholly laminar, whilst $n = 0.5$ would imply wholly turbulent flow. Often the Q_{50} value is converted to an air change rate at 50 Pa, or N_{50}. This is done quite simply by the relationship:

$$N_{50} = \frac{Q_{50}}{V}, \tag{4.6}$$

where V is the volume of the building (m^3).

The curve-fit method may also be used to extrapolate to a value of Q_{50}, if it is not possible to reach an applied pressure differential of 50 Pa. The use of extrapolation is not really ideal, as the result obtained will be sensitive to errors in measured data at lower applied pressure differences; in other words, in the range where errors are likely to take place. For the majority of dwellings, a pressure differential of 50 Pa should be readily attainable, and therefore the need for the use of extrapolation should not arise. This may well prove not to be the case with larger buildings.

Using the results of a fan pressurisation test, it is possible to estimate the total area of the cracks and openings in the building envelope by calculating the equivalent leakage area (ELA). ELA is defined as the area of sharp-edged orifice (the type of opening offering the least resistance to air flow), which would pass the same volume flow as the test building at the same applied pressure difference. It should be borne in mind that ELA is pressure difference dependent, as the area of some of the crackage in the building envelope will vary with applied pressure difference. It is therefore important that when a value of ELA is quoted, the applied pressure difference to which the ELA relates must be also be given.

A common way of calculating ELA[16] is by using the relationship:

$$\mathrm{ELA} = \frac{Q}{C_\mathrm{d}}\left[\frac{2\Delta P}{\rho}\right]^{1/2}, \tag{4.7}$$

where ELA is the equivalent leakage area (m^2), Q is the volumetric air-flow rate (m^3/s), C_d is the discharge co-efficient (set at an overall value of 0.6), ρ is the density of air (kg/m^3) and ΔP is the applied pressure difference (Pa).

Since ELA is merely a representation of the overall crackage area, it is therefore independent of both the building volume and surface area of the external building envelope. In order to be able to make comparisons between one or more buildings, it would be necessary to normalise the ELA. There are two commonly quoted parameters of this type. Normalised leakage area (NLA) is given by:

$$\mathrm{NLA} = \frac{\mathrm{ELA}}{\text{surface area of building envelope}}. \tag{4.8}$$

Specific leakage area (SLA) is given by:

$$\mathrm{SLA} = \frac{\mathrm{ELA}}{\text{floor area of the building}}. \tag{4.9}$$

Stephen and Webb[17] make the important point that when a pressurisation test is performed on a property such as a semi-detached or terraced house, a proportion of the overall measured air leakage rate will in fact not be to the outside, but instead will be to or from the adjacent property or properties via the party wall or walls. This is a difficult issue to resolve. Techniques have been developed which will enable only the air leakage through the external envelope to be determined. One way of doing this would be to pressurise or depressurise the adjacent property or properties so that zero net pressure difference(s) existed across party wall(s), as shown in Figure 4.7. This would mean that the measured air flow at a given applied pressure differential within the test property will be attributable to leakage paths through the external envelope alone. This idea has been applied to the measurement of air leakage through the walls of a range of dwellings.[18] The drawbacks of the technique are the extra equipment needed for field measurements, together with the greater complexity of the

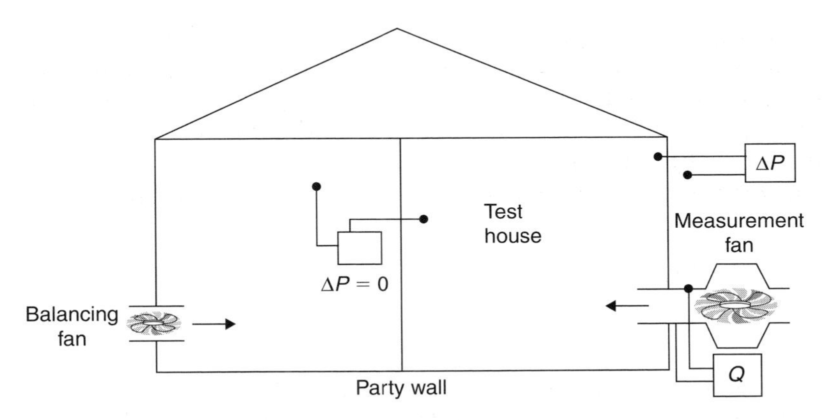

Figure 4.7 Pressure equalisation across party wall

test procedure. A much simpler method would be to use plastic sheeting to seal out surfaces of party walls. This approach could be used to determine the air leakage through any surface of the building envelope. The difficulties associated with this approach are the disruption caused within occupied dwellings, and the risk of damage to decorative finishes and possible furniture and ornaments.

Fan pressurisation can also be used as a means of locating leaks. For example, it would be very difficult, if not impossible, to find the minute crackage paths that might be found within masonry. However, it is relatively easy to identify the location of more substantial cracks such as those around ill-fitting window frames. The use of a smoke source such as a smoke pencil or an oil burning smoke generator can find such cracks. If the equipment and expertise are available, then the infrared thermography will provide an even more sensitive means of locating leakage routes. In fact, infrared thermography is so sensitive that it will sometimes detect air infiltration without the use of pressurisation.

The BRE have carried out extensive number of fan-pressurisation measurements on housing in the UK. They have found a wide range of air leakage rate, going in excess of 35 ach in some cases. The average value for the properties investigated seems to be about 12 ach. The recommendations for good practice in detailing contained within the 1990 Approved Document L of the Building Regulations for England and Wales[19] are generally accepted as being intended to result in a Q_{50} value of about 10 ach for newly built premises.

The Q_{50} value is intended to represent a value of air leakage at a pressure differential at which wind and temperature effects are negated. The question arises: how does the Q_{50} value relate to real conditions?

Typical pressures acting on a house are likely to average about 10 Pa. There is a very convenient rule of thumb from which an estimate of the background air infitration of the building can be derived. Quite simply, the Q_{50} value is divided by 20. This simple rule provides quite a reliable figure, but becomes less reliable as the building becomes taller.

4.6 AC pressurisation

Despite the treatments that can be given to the results of fan-pressurisation tests, the one piece of data that cannot be deduced is an accurate measure of the actual crackage area in the building envelope. AC pressurisation is intended to yield this data. Instead of applying a steady-pressure differential to the building envelope, a sinusoidal pressure wave is generated, either by means of a loudspeaker or by drum and piston arrangement.[20] The response of the building envelope to the pressure wave is analysed to yield information about the total crackage area. The equipment needed for AC-pressurisation measurements is barely portable, and highly skilled personnel are needed to make the measurements and analyse the data. For these reasons the technique is not likely to achieve widespread use, despite its undoubted value as a means of determining leakage areas at low-pressurisation differences.

References

1 DL Bohac. MSc thesis, Princeton University, 1986.
2 MDAES Perera, R Walker, JJ Prior. Simplified technique for measuring infiltration and ventilation rates in large and complex buildings: protocol and measurements. *8th AIVC Conference, Ventilation Technology – Research and Application*, Federal Republic of Germany, 21–24 September, 1987, Supplement to Proceedings, pp. 29–46.
3 RN Dietz, RW Goodrich, EA Cote, RF Wieser ASTM STP 904, pp. 203–264, 1986.
4 RE Edwards. Measurement of air change rates in wall cavities. *TRADA Report*, 1988.
5 RE Edwards, R Hartless. Measurements of subfloor ventilation rates – comparison with BREVENT predictions. *11th AIVC Conference, Ventilation System Performance*, September 1990, Belgirate, Italy, March 1990, Vol. 2, pp. 1–16.
6 C Irwin, RE Edwards. Airflow measurement between three connected cells. BSE and RT, No. 8, 1987, pp. 91–96, 1987.
7 C Irwin, RE Edwards. A comparison of different methods of calculating interzonal airflows by multiple tracer gas decay tests. *10th AIVC Conference*, Espoo, Finland, September 1989, Vol. 2, pp. 57–70, 1990.
8 RK Stephen, BC Webb. N148/85, Building Research Establishment, 1985.
9 BS1042, Section 1.1, 1980; Sections 1.2 and 1.4, 1984, British Standards Institution, London.
10 BS848 Part 1 1980, British Standards Institution, London.

11 BRE Fan Pressurisation Design and Calibration Information Paper, 1985.
12 RK Stephen, BC Webb. N148/85, Building Research Establishment, p. 11, 1985.
13 *Ibid.*, p. 7.
14 *Ibid.*, p. 15.
15 *Ibid.*, pp. 12–13.
16 *Air Exchange Rate and Airtightness Measurement Techniques – An Applications Guide*, pp. 12–13, AIVC, 1988.
17 RK Stephen, BC Webb. N148/85, Building Research Establishment, pp. 15–16, 1985.
18 JT Reardon, AK Kim, CY Shaw. *ASHRAE Transactions* 1987, 93(Part 2).
19 Approved Document L, Building Regulations for England and Wales, HMSO, 1985.
20 MP Modera, MH Sherman. *ASHRAE Transactions* 1984, 91(Part 2).

5

Ventilation Strategies for Dwellings

5.1 Trickle ventilators and other purpose-provided openings

The use of purpose-provided openings for providing ventilation is primarily intended to provide a fairly steady background air flow. This air flow can be controlled if required, perhaps on occasions when the incident wind is strong. A requirement for the provision of purpose-provided openings has been the part of the Building Regulations for England and Wales for the best part of two decades.

There are several possible means of introducing air for background ventilation. The most popular means of achieving this is via the use of trickle ventilators (Figure 5.1). These are incorporated into the frame of a window. There are a wide range of possible configurations, but the basic principle is the same. They can be made from a range of materials. Aluminium is popular, but some models are also available in plastic: the choice of material may be influenced by the actual material used to make the window frame itself. Modern unplasticised polyvinyl chloride (uPVC) window units are supplied with trickle ventilators readily fitted according to demand. In timber windows, an inexpensive trickle ventilator opening may simply be provided by cutting a slot in the window frame. The only extra component required used to be an internal open/close slider, which is simply screwed into the window frame. The current version of Approved Document F requires the provision of an integral insect-proof mesh.[1] This means that nowadays, trickle ventilators usually have an external cover with the insect screen built in. It is also a requirement that all purpose-provided ventilators shall be infinitely controllable. This requirement sounds rather

Figure 5.1 Trickle ventilator

demanding, but in fact can be easily achieved by the use of a sliding shutter; that is, in reality almost all types of adjustable trickle ventilators are "infinitely controllable". There seems to be no recorded evidence in the literature regarding the durability of these sliding shutter arrangements. However, the personal experience of the author is that approximately 50% of the sliding shutters fitted on the windows of a house built in 1986 were inoperable by 1999. This must be a concern for the medium- to long-term performance of such trickle ventilator control arrangements.

There is a reasonable amount of research literature and advice on applications for trickle ventilators. Some of these documents date back to the early 1980s. Jones and O'Sullivan[2] describe an investigation of the effects of trickle ventilators on the occurrence of condensation and the distribution of ventilation air through dwellings. The properties in question were considered to be of low-energy design with respect to the prevailing practice in 1985.

Out of the selected groups of 32 trial properties, 17 were fitted with trickle ventilators. Jones and O'Sullivan found that the whole-house ventilation rates in dwellings without trickle ventilators were seen to be satisfactory at about 0.5 air changes per hour. However, the consideration of conditions within individual rooms revealed that bedrooms and living rooms had ventilation rates that were much lower than the mean for whole dwellings. In some of the properties, problems classified as "serious condensation problems" were observed. In contrast, the presence of trickle ventilators in dwellings resulted in better distribution of air and more uniform air change rates. Consequently, a reduction in the number of condensation problems was noted in these latter properties.

The Energy Efficiency Office[3] describes a Building Research Energy Conservation Unit (BRECSU)-sponsored demonstration project that was intended to show the influence of trickle ventilators on condensation risk and odour problems. The results of the work showed that the use of trickle ventilators reduced both the incidence of odour problems and condensation, with the added advantage that they did not cause any significant increase in energy consumption. However, it was also noted that during the periods of high-pollutant (usually water vapour) generation, the ventilation provided by the trickle ventilators was inadequate. This was probably one of the first traceable acknowledgements in the literature that extra ventilation would be needed in addition to the trickle ventilators at certain times of increased water vapour production.

A much more detailed study of domestic ventilation strategies was carried out in New Zealand by Bassett.[4] The work centred around the use of a multi-zone air movement model as a means of predicting the influence of the relative proportions of purpose-provided ventilation openings and passive stack ventilation (PSV) on overall ventilation rates, with consideration being given to the effects of dwelling airtightness on the ventilation rates observed.

The results obtained show that ventilation rates of between 0.5 and 1.0 air changes per hour could be achieved when ventilation openings were used in conjunction with PSV ducts. The effect of the combined systems was more pronounced when installed in airtight houses. On the basis of the results obtained, Bassett suggests a set of sizing criteria for the ventilation openings. These are based on a linear relationship between the extra ventilation rate induced by the vents and the airtightness of the dwellings in question. In a subsequent piece of work, Bassett[5] describes the experimental results for four different sizes of window-mounted ventilation openings installed in three different unoccupied test houses. The sizes of the ventilation openings were calculated based on the criteria described in reference[4]. On this basis, ventilation openings between 21,000 and 110,000 mm^2 total open area were installed in the test properties. It must be said that these opening sizes are very large in comparison to the typical UK trickle ventilator open area of between 4000 and 8000 mm^2 depending on the room under consideration (readers should refer to Chapter 7 for more details of the actual regulatory requirement). The measured air-flow rates through the ventilation openings agreed with predicted values to a similar level of accuracy as for the measured whole-house ventilation rates with the predicted ventilation rates.

Despite their comparatively long-standing use in dwellings in the UK, research relating to trickle ventilators continues to the present day. For example, Ridley *et al.*[6] describe an investigation into the impact of the replacement of windows on background air infiltration within the dwellings. The study is based on the use of a simple model of air infiltration that allows the prediction of reduction in air infiltration within a dwelling given the results of laboratory tests on window units. Predictions are presented for new window units, both with and without controllable background ventilation. The main conclusion of this work is that the replacement of old windows could reduce ventilation rates to an unacceptably low level unless appropriately sized controllable background ventilation openings are fitted at the same time. In most of the cases this would mean the provision of trickle ventilator openings fitted within the window frames.

5.2 Passive stack ventilation

In the past, there was a tendency to use the term "passive ventilation" as an alternative to "natural ventilation". As a result, there was confusion between the former term and PSV. Fortunately, the term "natural ventilation" has now become the accepted terminology when describing about the ventilation strategies relying on wind and temperature differences. PSV refers specifically to the extract ventilation of spaces by ducted flows produced by wind-induced pressure differences and the so-called "stack effect". The application of PSV is generally limited to water closets (WCs) and rooms in which moisture-generation rates are likely to be high; that is, kitchens, bathrooms and utility rooms. For brevity, it is often called PSV, and this is the abbreviation that will be used in this text.

The principles underpinning the operation of PSV are not new; indeed, it has been referred to on more than one occasion as a refinement of the chimney. Of course, the chimney is many hundreds of years old, and its purpose was to remove smoke from fires. The concept of the use of ducts for extraction of stale air within hospitals is encountered in the UK at least as early as Victorian times. Further back in history, the use of wind towers in buildings in desert areas is one of the several commonly encountered design features. Their use dates back even longer than the chimney. The increasing popularity of PSV is viewed with some amusement by those who recall the days of solid fuel fires before gas central heating gained its mass appeal. The comparison of chimneys with PSV ducts should not be taken too literally. The old-fashioned chimney stack has much greater cross-sectional area than the typical PSV duct, the former involves the combustion process, and in any case the internal–external temperature differences associated with PSV operation are much smaller.

The first reference in the literature to a PSV-like systems in dwellings seems to be in an issue of the *Timber Trades Journal* of the late 1880s. Clearly the idea did not make much headway, although ducted extract systems were (as has already been mentioned) occasionally used in larger buildings such as schools and hospitals. In the modern era, PSV has been widely used in mainland Europe for about three decades. To the knowledge of the author, at least two companies have been selling system components for over 15 years, and in one case for over 20 years. At different times, the products of both of these firms have been imported into the UK, and most definitely prior to the changes contained within the 1995 version of Approved Document for England and Wales. Most sales probably arose as a result of the mention of PSV within the 1989 version of BS 5250. Even so, commercial interest was rather limited.

From the perspective of the typical UK house builder, extract fans provided a simple, inexpensive means of compliance with the Building Regulations, and there was little perceived benefit in deviating from this design solution.

Some international studies were carried out both prior to the increase in interest within the UK and within the same time frame. For example, Shaw and Kim[7] describe a series of air change rate measurements in a two-storey detached house. Five passive ventilation system configurations were investigated. The design options used might appear to be unusual from the UK perspective, but are fairly common in the USA. The options were:

- an intake vent located in the basement wall;
- a 10-cm diameter outdoor air supply ducted to the return duct of the existing forced air-heating system within the house;
- a 12.7-cm diameter exhaust stack extending from the basement to the roof;
- two combinations of the supply systems and the exhaust stack.

The best air change rates were found to result from the use of the outdoor supply duct with the exhaust stack from basement to roof. When the incident wind velocity was less than 30 km/h (8.3 m/s), it was noted that the stack effect is providing the principal driving force for the extraction of air via the exhaust duct. The house was also fitted with a fan supplying air to the boiler. (Note to readers – in American papers, the boiler is usually referred to as the furnace). When this fan was in operation, the indoor location of the vertical exhaust stacks did not have much effect on the overall house air change rate. This would perhaps suggest that the "furnace" supply fan was interfering with the overall performance of the ducts. However, Shaw and Kim suggest that even when the fan was running, the location of the exhaust stack could still influence the efficiency of mixing of outdoor air entering the house.

Consideration of this complex piece of work by Shaw and Kim leads to the conclusion that whilst what is presented is very interesting, it is most difficult to extract the elements of the finding that could be said to have any direct relationship to the circumstances that would be encountered within houses in the UK. Fan-powered air supply to combustion appliances has only come into vogue relatively recently in the UK. The notion of providing PSV to a cellar would be regarded as the most unusual indeed, but in theory could happen if a treatable room was located in the cellar.

Bergsoe[8] studied ventilation rates and humidity conditions within a group of approximately 150 detached houses in Denmark. The aim of the work was to establish installation and operation guidelines for the use of PSV systems within the Danish-housing stock. This is interesting given the well-established use of domestic PSV systems in Denmark, which to the author's knowledge was very common at least as long ago as the early 1980s. The air change rates measured by the use of the perfluorocarbon tracer (PFT) tube technique (it will be remembered from Chapter 4 that the tracer gas technique is the best suited for long-term average measurements of air change rates) within the test properties were regarded by Bergsoe as being low, being on average 0.35 air changes per hour. This is only slightly less than some of the results observed in UK dwellings of perceived good airtightness (which will be described in detail later in this chapter). Consideration of building envelope airtightness did not appear to be within the remit of the study, although given the level of importance provided to airtightness issues, it might be reasonably assumed that airtightness levels would be good.

In more than 80% of the properties in the test group, the measured mean air change rate was less than 0.5 air changes per hour recommended for Danish properties at the time. The mean relative humidity in living rooms was 45%, whilst in bedrooms the mean value was 53%. In addition to the measurement programme, an occupant questionnaire was administered within a sample of 2100 dwellings. Despite the low measured air change rates, it was found that the occupants of approximately 97% of the dwellings had perceptions of the air quality within their dwellings that amounted to being content.

The findings of this study are worthy of deeper consideration. They suggest to the author that the low ventilation rates are more to do with the airtightness of the dwellings and a lack of understanding of the relationship between airtightness and natural ventilation provision, rather than any actual fundamental flaw within the basic concept of the use of PSV.

The performance of PSV did not seem to have been the subject of a great deal of research in the early days in the UK. It was only in the early 1980s that an interest in PSV came about in the UK. One of the earliest references to the performance of PSV is due to Hardy of Newcastle University.[9] The research was carried out on behalf of the Northern Housing Consortium, which consisted predominantly of the representatives of local authorities in the north-east. It must be remembered (perhaps even fondly remembered!) that in those days, the vast majority of social housing stock was under the control of local authorities, and indeed was referred to simply as "council housing". Many Housing Departments either had a thriving Technical Department, or else could fall back on the services of a large, highly competent Architects Department.

The results of the study as reported by Hardy were probably not on reflection very encouraging from the point of view of promoting the use of PSV. The study itself did not make use of a sizeable group of test properties. This is always an issue in site trials, but need not in itself prove to be a fatal flaw. The principal difficulty with the Northern Housing Consortium study was the sizing of the ductwork itself. For some reason (or perhaps reasons) that has not been cited in the publicly available literature, the systems were designed around the use of 2-in. (50 mm) diameter rigid circular plastic ductwork. Whatever the reason for doing this, the fact remains that this choice was in itself highly detrimental to the prospects for the success of PSV. Such a small diameter extract duct offers a high resistance to the flow of air, even at the relatively low range of extraction rates encountered in PSV systems. Indeed, its selection could probably have been ruled out simply by looking at either the The Chartered Institution of Building Service Engineers or American Society of Heating Refrigerating and Air Conditioning Engineers (CIBSE/ASHRAE) guides.

The undesirability of using 50-mm ductwork was borne out by the results obtained. These were frankly unattractive. It is probably fair to say that the main reason for citing the Northern Housing Consortium study is to provide an early example of how not to use PSV systems in the dwellings. In the meantime, regardless of the experimental results published by Hardy, PSV systems continued to be used with success in mainland Europe. It was only a matter of time before a more convincing study was carried out.

The next relevant research work centred around the evaluation of the possible use of PSV systems as a means of ventilating high thermal performance timber-framed properties. The work was performed primarily by Gaze and Johnson, and was mainly funded by a partnership between the Timber Research and Development Association (TRADA) and Pilkingtons Research and Development. Once again, these were organisations that at the time had a significant stake within building research. TRADA had amongst its many research briefs for the timber industry and its main customers a requirement to provide the technical information supporting the use of timber-framed construction within dwellings. Interested readers should refer to TRADAs excellent design manual for timber-framed housing.[10] This publication is now close to 10 years old, but is nonetheless still well worth looking at. Pilkingtons were the keen collaborators in research projects in this area, as increased construction of timber-framed housing offered potential for good sales of glass-fibre insulation.

The measurements carried out by Gaze *et al.*[11] were performed in a sample group of four test houses on a new estate in Southampton that were built by Laing Construction. (Laings also had a thriving research and development unit at the time.) The sample properties were airtight in comparison to the typical new-built property of the time. The experimental methodology used involved the measurement of the variation of air change rates within the properties, both with internal–external temperature difference and the prevailing wind conditions. Tracer-gas measurements were carried out in order to determine air change rates by the decay method (for further details refer to Chapter 4).

It was found that when the houses were occupied, the PSV ducts gave consistent background ventilation rates, ranging between 0.3 and 0.6 air changes per hour with a mean value of 0.45 air changes per hour. These values were considered to be adequate for ventilation purposes, and the validity of this conclusion was underlined by the reactions of the dwelling occupants to the PSV systems. They noted that cooking smells and associated steam cleared away quickly, there were no reports of musty odours in bathrooms and, very interestingly, tobacco odours in living rooms dissipated overnight. This latter finding might seem to be surprising, particularly to non-smokers whose houses have been visited by smokers.

It was observed by Gaze *et al.* that "window vents", which are presumed by the author to be trickle ventilators, gave occupants the option of increasing ventilation rates. This form of wording would seem to imply that the trickle vents were not normally open. This, in combination with the relative airtightness of the houses, would seem to explain the relatively small spread of measured air change rates. On the basis of the measurements reported, the PSV ducts were not considered to constitute a source of excessive extraction during periods of high-incident wind velocity; neither were any instances of flow reversal reported. The PSV ducts were described by Gaze *et al.* as being "self-throttling". In the light of later work, this can now be seen to be a consequence of inadequacy of air supply to feed outflow through the ducts. Gaze *et al.* make the recommendation that in less airtight properties, draught stripping would be necessary in order to prevent excessive ventilation.

In a later publication, Gaze[12] gives further consideration to the issue of excessive extraction through PSV ducts. In order to minimise the risk of overextraction, he makes the recommendation that PSV ductwork should be decoupled at a point just below the roof terminal. This measure would certainly have the effect of throttling the air flow through the systems. However, with hindsight, it is now understood that such a decoupling strategy would have the most definite adverse side effect of permitting the passage of moist air from kitchens and bathrooms directly into the roof space. In view of the potential problems with condensation in pitched roof spaces described in Chapter 6, this would most definitely be undesirable. The use of decoupling has not been pursued in the subsequent research, and does not appear to have been utilised in many installations. On balance this is a very good thing.

On reflection, Gaze made his recommendations on the basis of performance data collected in properties that were much more airtight than most of the present-day housing stock, and they provided a solution which failed to recognise that the main performance issue was the inadequate provision of openings to feed the extraction of air via ducts. Given that one of the main perceived advantages of timber-framed construction was increased energy efficiency; the logic behind sacrificing some of the airtightness of the timber-framed building envelope in order to increase the rate of extraction would not have been appreciated at the time. Ironically, the relatively steady extract rates measured during the TRADA studies would nowadays probably be construed as a major plus. One of the major criticisms of PSV levelled by its critics is the lack of controllability that may be achieved. Perhaps, still there will be lessons to be learnt from this work as standards of airtightness for dwellings are progressively improved.

Another piece of research, published just 2 years after that of Gaze *et al.*, adopted a similar type of experimental strategy; in that its objective was to look at the effects of PSV ducts from the point of view of their influence on the actual air change rates in kitchens and bathrooms. As one of the researchers involved, the author recalls that this approach was adopted with the intention of producing some performance data for PSV systems that could in principle be fairly compared with those of Gaze *et al.* Edwards and Irwin[13] describe a programme of field measurements which sought to demonstrate the influence of duct and roof terminal on system performance. The field measurements were carried out in two very different types of property. The first was a highly energy-efficient, high-specification new-built property; this property is of timber-framed construction in keeping with best practice as recommended by TRADA.[10] As a consequence of the construction, the house was very airtight. By the way of complete contrast, the second house was a social dwelling under the control of a local authority. The house was awaiting refurbishment, and whilst not exactly derelict, was far from being in a good state of repair, particularly with regard to the state of window- and doorframes. Consequently, the house was very leaky. Both the properties were unoccupied when the measurements were carried out. Decay rate tracer-gas measurements were carried out in order to demonstrate the influence of internal–external temperature difference on PSV extraction rates.

The results obtained demonstrated noticeable differences in system performance between the two properties. Extraction rates through the ducts in the social dwelling were consistently higher than in the timber-framed house when only the trickle ventilators were open. Indeed, the results obtained within the timber-framed house suggest that the purpose-provided openings were inadequate for the provision of a flow of air into the property for the purpose of optimising PSV extraction rates. In other words, the PSV ducts were being throttled back as a result of inadequate air supply. In order to increase the air flow in the ducts to a level comparable with the social dwelling, it was necessary to resort to opening windows. This is clearly an undesirable state of affairs. The results themselves are entirely consistent with those quoted by Gaze *et al.*[11,12]

By the way of contrast, the PSV ducts in the social house were clearly shown to provide a significant enhancement to an already high background air change rate. Whilst the results for the social house serve to give weight to assertion of Gaze *et al.* that PSV should be used in conjunction with draught stripping in properties that are not airtight, the results obtained for the timber-framed house serve to underline the obverse, namely that PSV systems will not perform correctly if the amount of air leakage in the building envelope is not present. This might take the form of fortuitous leakage through the fabric or else through purpose-provided ventilators; if the amount present is not high enough, then PSV extraction rates will suffer.

Although not actually mentioned in reference[13], subsequent reflection on the results obtained for the timber-framed property led the authors to the conclusion that there was a significant shortfall in open area for air leakage in relation to the cross-sectional area of the passive stack ducts used in both bathrooms and kitchens. Interestingly, the open areas required for trickle ventilation within the most recent version of Approved Document F for England and Wales would have all but redressed the imbalance (see Chapter 8 for more information).

Irwin and Edwards used the measured air change rates to calculate the likely rates of moisture extraction during occupation using a methodology due to Meyringer.[14] They also compared the measured air change rate attributable to the PSV systems with values calculated by means of a simple theoretical approach, and concluded that the two are in reasonable agreement. As a final footnote to this piece of work, Irwin and Edwards observed flow reversal in ridge tile ventilators ventilating the roof space of the timber-framed property. This will be discussed further in Chapter 7.

From the late 1980s until the mid-1990s, a lot of research was carried out within the UK regarding the use of PSV. The works of Gaze *et al.* and others had clearly caught the eye of the legislators and other commercial interests, even though the sum total of performance data was not sufficiently comprehensive so as to provide the evidence base necessary to prove that PSV could safely be used in the UK housing. Once again with the privilege of hindsight, the main deficiency of the data collected was its principal focus on housing of an airtightness that was much higher than would be found in typical new-built properties of the time. Clearly some more studies on more representative properties are needed.

The Building Research Establishment (BRE) was heavily involved in the work. Initial work by Tull in a test house at the BREs' site at Garston (and seemingly not published) was inconclusive. Undeterred by this, the BRE persisted with the research. One of the first published studies was due to Uglow and Stephens.[15] Uglow also performed measurements in a test house at Garston. The work involved measurements on two different duct configurations. The first duct system was straight, leading from the kitchen of the test house to the ridge and terminating just above the ridge line. The second system also led from kitchen to ridge, but instead of being a straight system, incorporated two 45° bends located within the loft space of the house. Two different diameters of ducts were used, namely 100 and 150 mm, and the tests were repeated for both smooth rigid plastic and flexible plastic spiral-wound ductwork. Air-flow rates through the duct systems were measured and compared with an internal–external temperature difference and wind conditions. This was a different approach to the tracer-gas methodology employed in the work of Gaze *et al.*[11,12], and Irwin and Edwards.[13] A range of different types of roof terminals were used during the course of the measurements, in order to see if they had any effect on the system performance.

The results obtained indicated that for the straight 150-mm systems, air-flow rates were almost twice those measured for the equivalent configuration of 100-mm ductwork, and only 50% higher than for the case of systems with bends. This is an entirely reasonable observation given the fact that the cross-sectional area of 100-mm duct is only 64% of that of 125-mm ductwork. The inclusion of two 45° bends was seen to reduce air flows by 50%. There seemed to be no significant difference in air flows between the flexible duct and the rigid duct. With hindsight, this was an important finding that was not fully analysed at the time.

After the work of Uglow and Stephens,[15] studies of PSV continued at the BRE. In order to observe the effectiveness of commonly found PSV duct configurations in minimising the risk of condensation, Parkins carried[16] out a series of air-flow rate measurements in PSV systems and related them to relative humidities within the test dwelling in question. The properties under investigation were occupied, and were fitted with PSV systems installed by a contractor. Initially, the bathroom PSV ducts were found to be performing poorly. Extraction rates were low; in fact, so low that no reconciliation could be made to the claimed system design. Inspection of the flexible ductwork in the roof space revealed that a large section of the duct-run had been laid parallel to the ceiling, and consequently there were very sharp bends in the system. An excessive length of duct had also been used. Apparently the contractor had not seen fit to cut the flexible duct supplied to length. The problem was made worse by the absence of any support for the duct. It was left free hanging between the ceiling extract and the roof terminal. When remedial measures were applied to the duct, a significant improvement in performance was noted. The relative humidity was seen to be below 70% for all but a very small proportion of time.

On the basis of the research, Parkins observed that there were several key aspects of system design that needed careful attention in order to ensure that optimum extraction rates are achieved. In particular, this study identified the importance of straightness of ductwork, and the position of the roof terminal as determinants of the extract rates that could be achieved through PSV stacks. A particularly significant statement made by Parkins relates to the use of tile ventilators as PSV system terminals. In the opinion of Parkins, the use of tile ventilators below ridge height should be avoided, as their use was often responsible for flow reversal. Whilst the results presented lend credence to this conclusion, the experimental base for the conclusions was very narrow, and was clearly not sufficient to justify a blanket prescription of the use of tile ventilators for this specific purpose. This issue is further addressed in Section 5.3.

It is as a result of the work carried out by the BRE that the idea of sizing PSV ducts according to the room to be ventilated and its location was mooted, not only with the purpose of ensuring adequate levels of extraction, but also to avoid excessive ventilation by using too large a diameter of duct. In terms of building services engineering practice, this approach was a reasonable one to take. The inadequacy of 100-mm ductwork for use in kitchens, in the opinion of the author, clearly established as a consequence of the BRE work.

Successful efforts were made to incorporate the prediction of passive stack duct extraction flows within the BREVENT single-cell air infiltration model. Cripps and Hartless[17] describe a study in which BREVENT predictions were compared with passive stack extract rates measured in the same test house used by Parkins.[16]

The sum total of available evidence, as detailed above, was allegedly considered by the Department of the Environment and the Building Regulations Advisory Committee prior to the introduction of the revised Approved Document F that came into force in 1990. Again allegedly, a decision was made not to include mention of PSV as a possible means of compliance because of reservations not only about the ability of PSV to give adequate extraction rates, but also the danger of draughts being caused. This may seem have been a cautious decision, but given the evidence available at the time, was probably the safe one to take.

Palmer *et al.*[18] presented the results of a study of measurements within properties in two low-cost housing schemes. One of the schemes involved the retrofitting of PSV ducts into local authority-controlled social housing. The properties involved had a record of condensation problems. The other scheme was an installation within a new-built property. The air change rates induced by the PSV ducts were shown to be of the order of 1 air change per hour. Palmer *et al.* found that the siting of the roof terminal was an important influence on overall system performance. They also reported that in cases where the terminals were sited at ridge height, very little reversal of air flow in the PSV ducts was noted.

It was appreciated at a fairly early stage of research that the influence of system terminals, both at ceiling extract and system discharge, would exert an influence on passive stack system extraction performance. If nothing else, the two terminals would probably represent the two biggest single contributions to the overall system resistance. The issue of ceiling terminals has probably been the easiest to resolve. The most popular design solution encountered by the author seems to be the type of circular ceiling terminal commonly known as a *register*, as used for extraction via mechanical systems (see Figure 5.2). The open areas of these terminals are adjustable for commissioning and system balancing purposes. When opened up to a large open area, ceiling registers are eminently suitable for use in PSV systems. The more thoughtful PSV installer might see the presence of the ceiling register as an opportunity to adjust the extraction rate through the duct in accordance with the airtightness of the building. However, there is little evidence to point to this happening in practice. The danger accompanying the use of adjustable registers is that they are susceptible to interference by occupants, who may well choose to close them right down if the mood takes them.

In comparison with ceiling extract terminals, the performance of roof terminals is much more complex. In the case of the roof terminal, there

Figure 5.2 Ceiling register

are other issues, apart from component resistance, that will have a bearing on the extraction rate via a duct system. In addition to this consideration, the aerodynamic design of the roof terminal will in itself make a contribution to the extraction performance, as will the location of the terminal itself. de Gids and den Ouden[19] reported the results of a study of pressure distributions around a model building, with the objective of identifying the most favourable location at which to terminate a PSV duct. The duct in question was a simple pipe discharge. The results showed that if the duct left the building at ridge height, then it would be sufficient to terminate the duct at a height of 0.5 m above the ridge itself. This location should guarantee that the discharge is always in the negative-pressure region above the roof.

The work of Welsh[20] is concerned with assessment of the performance characteristics of a wide range of commonly used system roof terminals. The study covered a wide range of terminal types, including the so-called Chinese hats, rotating cowls, H-pot terminals and gas flue terminals. Welsh carried out a series of tests under laboratory conditions that centred around pressure losses associated with each type of terminal, together with performance under different wind conditions, with the objective of obtaining a ranking of the overall performance of the terminals. On the basis of the results obtained, Welsh divides the terminals into three distinct groups:

- The terminals in the *first group* included a 110-mm gas flue and one type of H-pot terminal. These devices were considered to have high-loss factors, which would presumably make them a poor choice for a passive stack terminal.

- The *second group* included one type of rotating cowl and one type of H-pot. The results demonstrated that these terminals were effective in inducing upward draughts within passive ducts.
- The *third group* comprises three terminals, the open pipe, the mushroom cap and the Chinese hat terminal, these were identified as being liable to cause flow reversal in passive stack systems.

The findings of this study have significance to the design of passive stack systems, as terminals with high-loss co-efficients and a tendency to cause flow reversal would be the unsound choices, given the relatively low-pressure differences driving flows. From the aesthetic viewpoint, however, many architects might baulk at the aesthetic implications of using H-pot terminals. Indeed, their use might be proscribed in preservation areas by planning departments. Cost conscious contractors would doubtlessly be hostile to the use of expensive rotating metal cowl terminals.

The results of Welsh's research were of great interest. However, at present it cannot be said that it has diverted many of those responsible for the design of passive stack systems away from lower performance yet cheaper and more popular options such as ridge tile terminals and tile ventilators. As far as most of the house builders are considered, price is all.

Finally, in the discussion of the features affecting the performance of passive stack systems, it is all too easy to forget that there is an unavoidable human influence. The amount of research into the behaviour of occupants has been rather limited in comparison to that related to system design, installation and performance, but has nonetheless been interesting. One good example, although not carried out in the UK, is the work of Van Dongen.[21] The major element of the study consisted of a questionnaire administered to the occupants of dwellings during periods of mild winter weather on which external air temperatures were of the order of 5°C. This was combined with ventilation rate measurements. The occupants lived in three groups of dwelling types: natural ventilation by passive stack only; mechanical extract ventilation, and balanced mechanical supply and extract systems.

The results obtained indicated that the type of ventilation opening and its ease of use determined the amount of ventilation within the individual dwellings, rather than the behaviour patterns of the occupants. In particular, ventilation rates were not influenced by the level of occupancy. However, interestingly a positive correlation was found between the amount of use of mechanical extract ventilation and the number of occupants per dwelling. It might be speculated that this was due to increased levels of usage within bathrooms and showers. Furthermore, it was discovered that there was an overall tendency to under-ventilate dwellings in all the three groups. The reasons for this are not fully explored.

5.3 Passive stack system design

PSV is specifically mentioned in the 1995 version of Approved Document F of the Building Regulations for England and Wales. PSV was brought into the Scottish Building Regulations in 1997. A good starting point for system design is BRE Information Paper IP13/94.[22] It is important to remember that this document gives only guidance on designing systems which comply with Part F, and that other systems may be deemed to comply if it can be demonstrated, if they do so, preferably by experimental data. The significance of this will become apparent later.

IP13/94 gives information on recommended diameters of duct (see Figure 5.3) and other aspects of perceived good practice in system design, which can be summarised as follows:

- Have no more than two bends in each system, with no internal angle greater than 45° (see Figure 5.4).
- Insulate all ductwork so as to avoid the risk of condensation within sections in the loft.
- Ensure that all ceiling and roof terminals have an open area at least equal to that of the cross-sectional area of the duct being used.
- Terminate systems at ridge height, either by a ridge tile terminal or by a tall pipe discharging at ridge height (see Figure 5.4).

Whilst providing a reliable basis for design, IP13/94 can be said to be the most cautious document, and it is probably appropriate that it should be so. However, here are several areas in which research has demonstrated that there are more flexible options that can be used. One of the most important of these is with respect to the choice of roof terminal. Prior to the publication of IP13/94 it had been a common practice to terminate PSV systems by means of a tile terminal similar in appearance to those which might be used for roof-space ventilation (see Figure 5.5) in circumstances where it was impractical to use a ridge-tile terminal, for example on a short ridge board where other terminals had already been fitted or where it would be impossible to run a duct to the ridge without using excessive bends. The tiles were usually sited not more than three rows of tiles below the ridge, although it must be acknowledged that in some

Room type	Duct diameter (millimetres)
Kitchen	125
Utility room	100
Bathroom (with or without WC)	100
WC	80

Figure 5.3 Recommended duct diameters from IP13/94

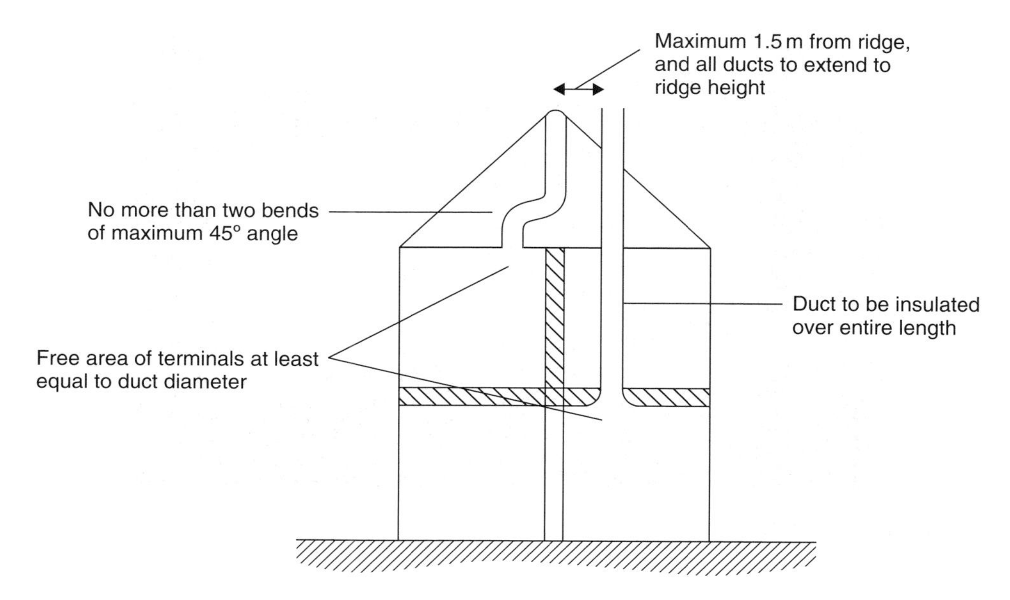

Figure 5.4 Recommended duct configuration

Figure 5.5 Tile ventilator terminals on roof

cases, problems had arisen with terminals being placed well down the roof pitch. A study of four test houses comparing the performances of PSV systems with different roof terminals concluded that providing that the terminal was not sited more than three rows below the ridge, the measured extraction rates were comparable with those expected for systems terminated at ridge height.[23] These findings have significantly extended the scope for the use of PSV within the housing stock.

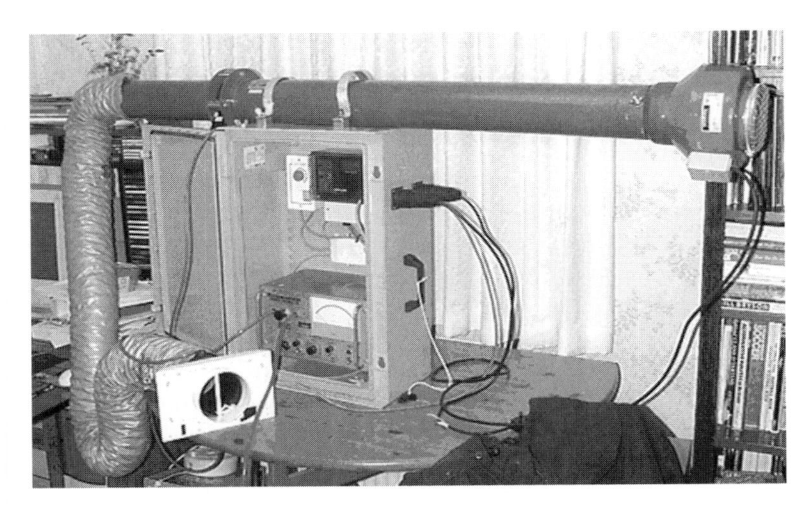

Figure 5.6 Portable duct testing kit

Another issue with regard to IP13/94 relates to bends in systems. It is undeniable that sharp bends or bends with other undesirable design features will cause a big drop in extract rates. This is particularly true for flexible ductwork. However, when properly radiused and secured bends are considered, a different picture emerges. Recent modelling work,[24] carried out using ASHRAE published data for bend resistance,[25] suggests that correctly radiused bends of as much as 75° will have little impact on the extract rates observed in PSV systems. The emphasis of this chapter is on the correct forming of the bends. In a subsequent piece of work, Edwards[26] presents the results of experimental pressure/air-flow measurements obtained by the use of a portable pressure-testing kit (see Figure 5.6). These results underline the conclusions reached in reference[25]. The author suggests the introduction of a simple performance criteria for PSV ducts, namely that at 50 m^3/h the total pressure drop across the duct should not exceed 3 Pa.

There is a postscript to the issue of bend-angle restrictions. In 1999, the Energy Efficiency Office published a guide to energy-efficient ventilation systems in dwellings.[27] Each of the main options is described and briefly discussed. In general, the document is a compilation of information and good practice prevailing at the time, and as a result little new information is added. Certainly, the reference list given at the end of the document supports this observation. Therefore, it is most interesting to note that the document recommends a maximum bend angle of not greater than 35° from the vertical; in other words, the suggested restriction on bend angles has been made more stringent. No reason is apparent for this change; that is, certainly no technical evidence is referred to. It is not clear whether the reference to 35° bend angles was even deliberate.

5.4 Comparison of means of extraction

PSV is intended to provide a continuous if variable means of extraction. Over typical day its total amount of air extraction will be approximately the same as a mechanical fan,[28] but the pattern of extraction will be very different. Due to the influence of stack effect and in particular the incident wind, PSV does not have the same level of controllability than mechanical systems. If systems are correctly designed, then variations in flow rates will not matter in the long term. The PSV extraction rate will increase with higher internal temperatures, which are co-incident with cooking and bathing, but a ramp increase in extraction will not be observed. This means that peaks of relative humidity in excess of 70% will be observed during the period of high-moisture production. However, such peaks are of little importance with respect to the overall satisfactory control of condensation provided by PSV systems, since they will be of relatively short duration if they occur, and will therefore not give rise to condensation and mould growth problems.

One definite advantage enjoyed by PSV is its low-maintenance requirement. Over and above cleaning of the ceiling terminal perhaps once per year, very little attention is required. Some concerns have been expressed about the danger of accumulation of grease deposits in ducts serving kitchen installations; but at the time of writing, no reports of actual problems have been published, and on this basis it may be reasonably assumed that regular cleaning of domestic PSV ducts should not be required.

In dwellings more than two storeys in height, attention must be paid to fire-safety considerations. In particular, ductwork must be of an appropriate fire rating, and heat-actuated dampers must be fitted where a duct passes through one fire compartment into the next. For more details on requirements for fire protection, the reader is advised to consult reference[29].

5.5 Improvements on the basic PSV system

The basic PSV concept has been shown to give suitable extraction rates on average basis. In many properties, particularly those where the airtightness of the building is not good, or else where the level of moisture production within the dwelling is not very high (perhaps where the property has only one occupant), the extraction rates through standard PSV systems may on occasion prove to be in excess of what is actually required. This will be particularly the case when wind velocities are high. The energy consumption of the dwelling in question will therefore be increased.

The overall acceptability of PSV systems could be improved if excessive extraction and energy consumption could be reduced. This could be done by reducing the rate of extraction at times when such a high level of ventilation is not required; in other words, by the use of a control system. In commercial buildings, use is made of several types of ventilation control. These include:

- *occupancy sensors*, which modulate ventilation rates according to the perceived level of occupancy within the building or its individual rooms, see for example the work of Francis and Edwards[30];
- *carbon dioxide sensors* or *air-quality sensors*;

In commercial buildings where extra ventilation provision is available, the extra provision is activated by *temperature sensors*.

Barthez and Soupault[31] describe a study of ventilation control using both carbon dioxide and relative humidity sensors, carried out in an occupied house in France. Two ventilation strategies were studied. A two-speed extract fan was used for mechanical ventilation, with the fan speed being modulated by the sensors. Similarly, modulation of natural ventilation was provided by using the sensors in conjunction with an exhaust duct damper. The objective of the ventilation control strategy was to achieve a relative humidity during occupancy of 60% at a temperature of between 18°C and 20°C, whilst not exceeding a carbon dioxide concentration of 800 ppm (which will be noted as 200 ppm lower than the generally accepted air-quality target for commercial buildings).

The monitoring results seem to show that whilst there is a strong correlation between carbon dioxide concentration and occupancy behaviour, the same cannot be said for relative humidity. In view of the sporadic nature of peak moisture production within dwellings, perhaps this might be expected. On this basis, Barthez and Soupault conclude that carbon dioxide is the most suitable parameter for the modulation of the ventilation systems. This is a most surprising conclusion given the nature of condensation problems within dwellings, and should be treated with a great degree of scepticism, according to the author. A daily mean air change due to either extract system of between 0.5 and 0.6 air changes per hour (in addition to uncontrolled ventilation through the fabric of about 0.2 air changes per hour) is shown to be adequate in keeping relative humidity and carbon dioxide concentration within the prescribed limits, except when cooking was taking place, when an increase in extraction to 0.8 air changes per hour was required.

In the domestic situation, most of the control sensors used in commercial buildings are likely to be of limited use. As has previously been explained in Chapter 3, the major indoor air pollutant within dwelling is water vapour, despite the findings of Barthez and Soupault.[31] It would therefore be sensible to base a control strategy for PSV systems on the amount of water vapour in the inside air. This is the principle for the commercial systems currently available. In theory, a system could be specified which used a relative humidity sensor to control a damper within a PSV duct. This is the approach that would be logical to the building services engineer. However, whilst such a control system might be entirely acceptable within the non-domestic situation, in a house, the associated costs and increased system sophistication would deter potential users.

The most successful control devices incorporate a relative humidity sensor and damper mechanism into the ceiling extract terminal of the PSV duct. The sensors used are mechanical in nature rather than electrical or electromechanical. They have direct linkage into a mechanical system for the modulation of the damper. A popular terminal is the Aereco "Hygro", as shown in Figure 5.7. The terminals are designed for use with 125-mm-diameter circular duct. The relative humidity sensor in this terminal is a strip of nylon. The length of the nylon increases with increasing relative humidity; for example, when extra water vapour is being generated in the space served by the passive stack duct, this allows the damper in the terminal to open, hence increasing the open area of the terminal and allowing a higher rate of extraction. When the relative humidity drops again, the nylon contracts, and the extract rate drops again. There is a facility to change the control range of the terminal. The normal setting, and the one that is appropriate for the majority of dwellings, is between 30% and 70% relative humidity. Purpose-provided wall and trickle-vent inlets are also available for use in conjunction with the Hygro ceiling extract terminal. In addition to its use on PSV systems, the Hygro-type terminal is eminently suitable for use with a ducted mechanical extract ventilation system. Indeed, the Aereco company offers a range of terminals and dedicated fan units in its product range. Test data suggests that energy savings of as much as 25%

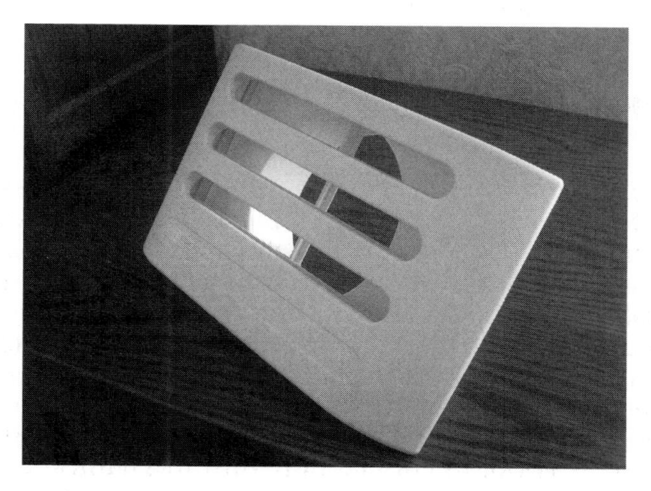

Figure 5.7 Aereco Hygro extract terminal

might be achieved by using Hygro terminals instead of non-modulated types.[32]

A similar system is offered by the Danevent company of Denmark. Instead of using a nylon strip, the terminals produced by this company use a thin membrane of beech-wood laminate. The wood swells as relative humidity increases, and the expansion in this case is constrained so as to produce a flexing of the membrane that causes the damper to open. As for the Hygro terminal, complementary air inlets are available.

The designs of humidity-controlled passive stack extract terminals are clever yet not complicated. Their use has been clearly shown to give energy savings. However, there are two issues that might lead to difficulties. Firstly, the use of the terminals is associated with higher internal relative humidities, as a consequence of reduced extraction rates. There is a risk that this might lead to an increased incidence of surface mould growth under the wrong conditions. Secondly, since the control of ventilation is based on the relative humidity within the space, humidity-controlled terminals may suffer from the same problem as relative-humidity-controlled mechanical extract fans; in that extract rates may remain higher than necessary, simply because a reduction in internal temperature causes an increase in relative humidity without any actual increase in the amount of water vapour contained in the internal air. This does not lead to such a problem as seen with simple fans without night setback, but could in theory still lead to higher air change rates at night than might be necessary. To the author's knowledge, there is currently no available variation of a passive terminal that has a night setback arrangement. Devices relying upon relative humidity as the sole control parameter carry the risk of giving over-extraction at night when internal temperatures fall. Devices using more complex control strategies (e.g. electronic sensors and motorised dampers) are likely to be rejected on the grounds of increased unit-cost and maintenance requirements.

5.6 Mechanical extract fans

Mechanical extract fans are a simple means of providing boost ventilation at periods of high-moisture production, and have been used for many years. The simplest types consist of nothing more than an electric motor, an impellor, a switch and a casing, as shown in Figure 5.8. Extract fans can be either wall or window mounted, and discharge directly to outside. Whilst fans can be mounted in double-glazed window units, this introduces a point of weakness into the unit in the form of another seal in the glazing system. As a result of the almost universal use of double glazing, now only window-mounted fans are used if absolutely necessary.

Recourse to ductwork in conjunction with mechanical extract fans is rarely needed, the most common exception being when the fan is packaged

Figure 5.8 Typical domestic extract fan

as part of a cooker hood assembly. In the case of the cooker hood, a choice of several volume extraction rates will be available by selection.

5.7 Assessment of extract fan performance

When assessed in capital cost terms, extract fans usually offer an attractive option. An extract fan will almost always be cheaper than any other means of providing direct ventilation from a zone of high moisture production. However, beyond this basic capital cost consideration, they can incur significant maintenance and replacement costs. Service lives may be reduced drastically if water or condensate finds its way into the motor. This may be a more common event than might be thought. In the same way that air supply and extract points are of concern in supply and extract systems, care must be taken when locating extract fans in order to avoid short circuiting of incoming fresh air directly to the fan. An example of this is shown in Figure 5.9. If the window is open, or the door is left ajar (ironically, this will often be done in an attempt to increase ventilation) outside air will short circuit directly to the fan without diluting any of the moisture being generated by the cooker. The fan is, therefore, not making any useful contribution to the control of condensation (or for that matter the combustion products, if the cooker is of the gas type). Similarly, there is a danger that all or some of the discharge from the fan may be re-entrained into the building if fans are not suitably located with respect to windows. During periods of operation, an extract fan causes a major change to the pressure distribution within the house. For example, air movements to the roof space may be reduced by as much as 30%.[33]

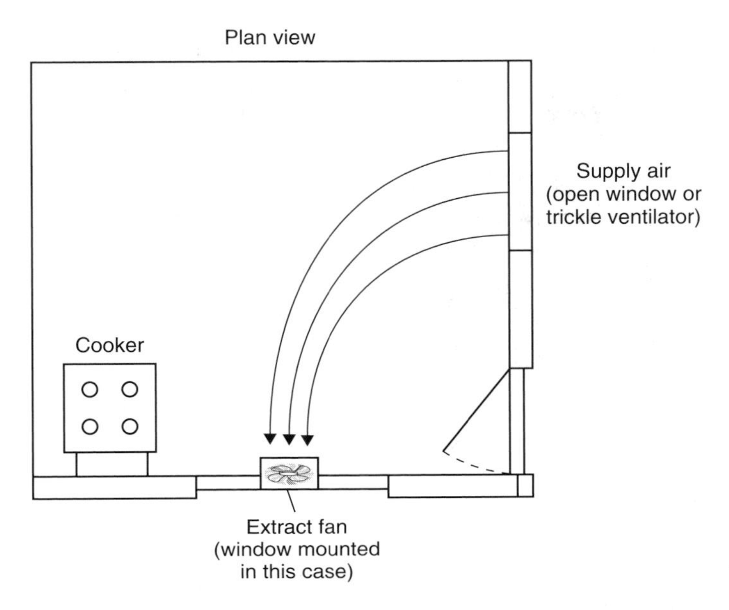

Figure 5.9 Extract fan short circuit

Fans can be noisy, and this may encourage occupants to switch them on. With suitable controls, some energy-efficiency improvements may be achieved. Fans controlled by relative humidity sensors alone should be avoided, as they may switch on during the night when internal temperatures drop. This problem can easily be overcome by the provision of a night overrun facility, but of course this may have a cost implication. Finally, an extract fan is an electromechanical device, and as such has a finite life. Fans within dwellings will at some stage need replacement if ventilation is to remain effective. If broken fans are left in place, then serious condensation problems may result.

5.8 Mechanical ventilation and heat recovery

Mechanical ventilation and heat recovery (MVHR) has been successfully used in ventilation systems within commercial premises on a regular basis. This is due to the fact that within high-flow-rate regimes, conditions are fully turbulent, and therefore the efficiency of heat transfer is good. This goes some way to counteracting the poor rates of heat transfer from solid surfaces to airstreams. There are several types of equipments available.

A typical domestic system is shown in Figure 5.10. The fans and heat exchangers are usually supplied in a form of a packaged unit. This unit will be usually housed in the roof space on account of its size. In some cases, restrictions on the size of the loft access hatch may make the retrofitting

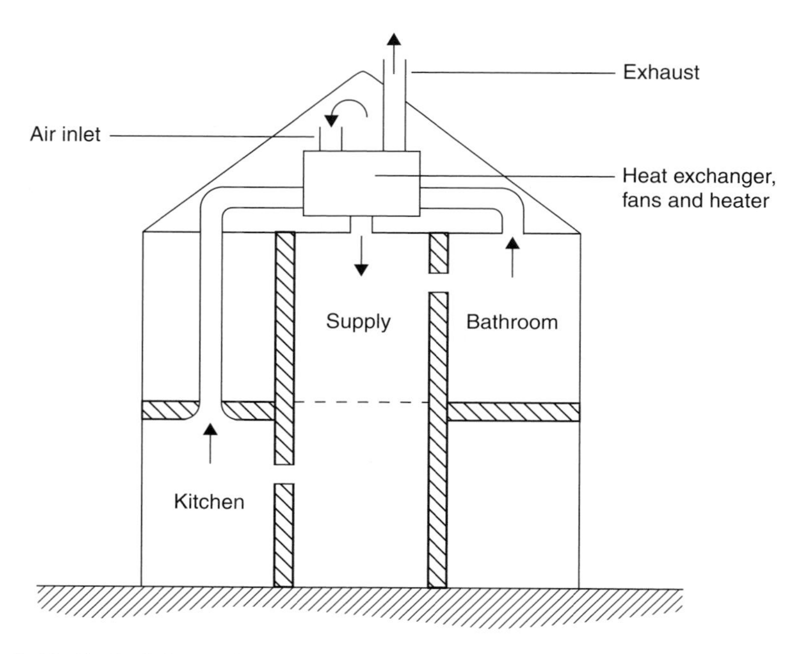

Figure 5.10 Typical domestic MVHR system

of MVHR problematic. A system of ducts of appropriate diameter is used to supply and extract air within the occupied space.

Within systems for individual dwellings, the likely air flows required for satisfactory ventilation mean that within the heat exchanger, whatever its type, it is probable that the air-flow regime will be in the transitional region. Heat transfer is the most efficient when the air-flow regime is fully turbulent. This issue has serious adverse consequences for the efficiency of heat transfer and hence for heat recovery. In addition, scope for controllability of air flow through the ventilation system is greatly diminished; since any reduction in flow rate beyond a critical point will cause a large drop in heat transfer efficiency (as shown in Figure 5.11). In practice, most units for dwellings are oversized, and the result of this is that excessive ventilation may occur.

Attempts to increase flows by the use of narrow air flow paths within the heat exchanger will result in an increase in fan power to overcome resistance and hence a decrease in coefficient of performance (COP). (The COP is defined as the ratio of amount or rate of energy recovery to the amount or rate of energy expenditure of the device recovering the energy. For a device to be of benefit, the COP must be greater than unity.)

The use of MVHR in multiple occupancy buildings such as flats would prove to be much less problematic, as a small number of high flow rate units could be used to service a large number of dwellings, and the heat transfer within heat exchangers would be much more efficient. In larger

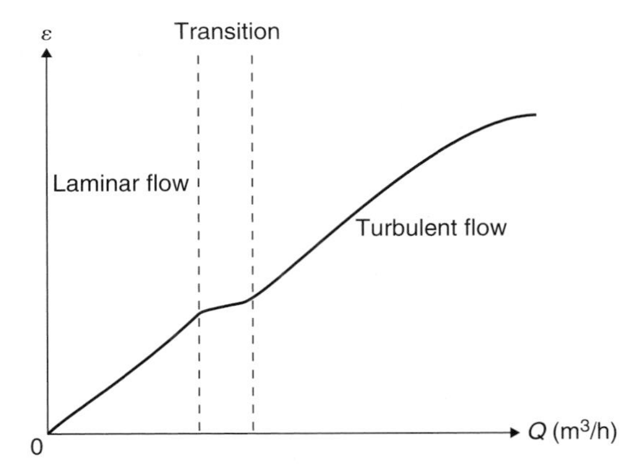

Figure 5.11 Efficiency of MVHR with changing air-flow rate

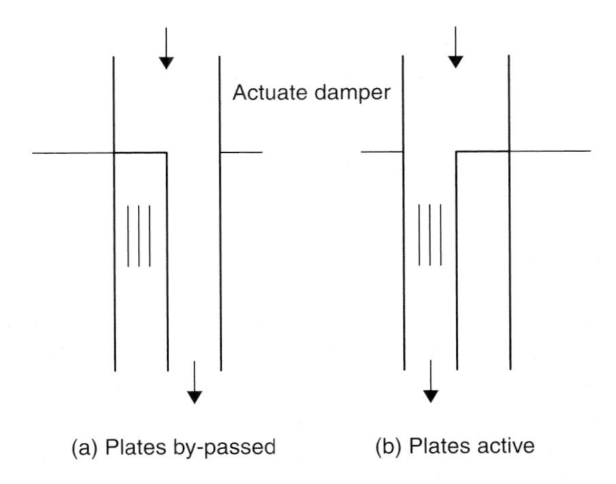

Figure 5.12 Face and by-pass arrangement

systems, some degree of controllability can be achieved by means of the use of "face and by-pass" arrangements, as shown in Figure 5.12. The arrangement has much similarity to that of modular boiler installations. This gives the capability of cutting off the air flow through some parts of the heat exchanger, whilst maintaining suitable air-flow rates through the open parts and hence not compromising the efficiency of heat exchange. A price is paid in the form of increased capital and maintenance costs.

Another serious drawback to MVHR is its interaction with air infiltration through the building fabric. An MVHR unit has no significant effect on the pressure regime within the dwelling. This means that the air flow through it is in addition to the infiltration through the fabric. This contrasts with the effects of extract-only systems such as PSV and extract fans, which reduce air infiltration as a result of induced depressurisation. For example, a PSV duct can reduce infiltration by as much as 25%.[34] It is most important to realise that a heat-recovery system can only recover heat from that fraction of the total ventilation air that actually passes through it. Therefore, it follows that within a typical UK dwelling, an MVHR system may only be able to recover heat through a much reduced proportion of the overall ventilation air. Overall COP can, therefore, be much lower than the measured bench COP for the unit used. The calculation of real COP values for heat-recovery systems is discussed in Section 5.10.

There are other considerations. Whole house MVHR units are still relatively expensive in relation to other alternatives, as they involve the installation of ducted ventilation systems. The use of ductwork results in a reduction in available space within the occupied space. The units themselves are quite large: so large, in fact, that it may well be the case that the loft hatch may be too small for the unit to pass through it. This severely limits the installation of units as a retrofit measure unless other major works are being carried out at the same time. Heat recovery units have associated maintenance costs which are likely to be high, and which will probably not be within the abilities of the householder, and throw away heat exchanger cartridges not withstanding. Failure to maintain a unit will result in a reduction in performance.

5.9 Variations on MVHR

Not all systems conform to the accepted model. Some use has been made of positive pressurisation systems. Such a system is shown in Figure 5.13. This type of system recovers heat from stale air displaced from the occupied space by means of a pressurised supply. Condensation risk is thus reduced, and good energy efficiency is achieved. This type of system has one obvious potentially serious disadvantage in that it encourages the permeation of warm moisture laden air into the building envelope. Whilst this might not pose much of a problem in traditional walls, there is a definite danger that in a highly insulated structure and, in particular, timber-framed buildings a risk of damaging interstitial condensation may be caused.

Of further concern is the issue from where the system actually draws its supply air. It might be expected that this air would be drawn from the roof space. However, if the upper floor of the dwelling in which the system is installed has significant crackage, poorly sealed service penetrations or an ill-fitting loft hatch (the same set of adverse fabric issues having a significant influence on roof-space condensation risk), then the unit will effectively be partially recycling warm air from inside the occupied space. Worse still, the flow of water vapour into the roof space may lead to condensation.

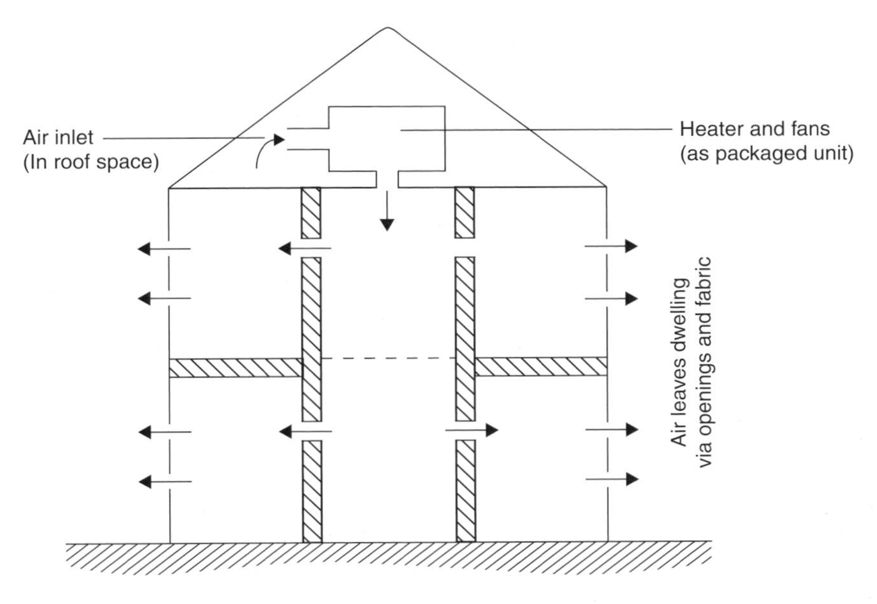

Figure 5.13 Positive-pressurisation ventilation system

Positive-pressure ventilation systems have been available in the UK for a good number of years; in fact, as long if not longer than PSV systems. Despite this, there is a curious dearth of published experimental data which provides evidence one way or another as to the actual usefulness of positive-pressure ventilation systems.

The limitations of standard centralised MVHR for dwelling have long been recognised. Attempts have been made to produce simpler solutions that make use of the principle of heat recovery without the costs associated with central systems. In particular, wall-mounted single-room units are available in a range of sizes. At the top of the range are units that give ventilation rates in excess of 500 m^3/h, such a unit would clearly be too large for the majority of dwellings. Other units are available which are capable of extracting from about 60 m^3/h at standard setting to a maximum of 200–225 m^3/h at boost setting, at an efficiency of between 60% and 75% depending on the extraction rate. Such a unit would be appropriate for use in a domestic kitchen. A third and rather smaller range of units are designed to give overall extraction rates of approximately 25 m^2/h in standard operating mode, with a boost facility to about three times this extraction rate. Heat-recovery efficiencies of between 50% and 60% are claimed by various manufacturers. Once again, the actual recovery efficiency of the process depends on the selected extraction rate.

Whilst in principle going a long way towards overcoming the problems associated with the use of centralised units, single-room MVHR will still be adversely affected by excessive envelope air leakage. Its use also multiplies up the need for maintenance and ultimately replacement, despite the simple and ingenious heat exchanger plate cartridges that are available. Unless householders regularly clean or replace the cartridges, the efficiency of heat recovery will be severely impaired. Complex and perhaps undesirable air-flow patterns may result if all installed units, which are not switched on at the same time. It should be selection of units, which is of great significance. For example, two units of the smallest size installed in a typical house would not provide a whole house extraction capability in line with the continuous 1 air change per hour required in Approved Document F. On the other hand, two of the intermediate units would achieve this, but the ventilation regime would not be "whole house" in the sense that no direct ventilation would be provided by the systems for rooms in which ventilation systems were not installed. The precise role of the units is, therefore, ambiguous, and their compliance or otherwise with the requirements of Part F could be a matter of some discussion. The manufacturers provide some clues by virtue of the fact to that they offer a range of control devices for use with the units, the most appropriate of which for the domestic situation are relative humidity sensors. This, in the opinion of the author, gives a clear indication that single room heat recovery units are intended for use on the following basis:

- to give background ventilation at the lowest extraction rate when a high rate is not required;
- to increase the extraction rate to a "boost" value when required (e.g. when cooking or bathing is taking place).

This regime is similar to a standard mechanical extraction fan, with the important difference being that the unit is used to provide background ventilation outside the periods of high moisture production. The most significant implication of this is that for much of the time, any units installed within a property will be handling a much reduced proportion of the overall ventilation air in the property and, therefore, the basis for heat recovery is very different at these times. If the units were left running on a continuous basis at the higher extraction rate, then for the majority of properties an excessive amount of ventilation air will be extracted from the property with a consequent increase in energy consumption for space heating.

Other difficulties arise as a result of the design of the units themselves. As it is very important on both space and aesthetic grounds that units are of as compact a design as possible, single room heat recovery devices have supply and extract louvres located very close to each other, both internally and externally. As a result, there is a risk of short circuiting of air, which may possibly lead to both inadequate ventilation inside the room, and also the re-entrainment of pollutants from the exhaust to the inlet. These problems might arise in addition to any reduction in the energy efficiency of heat recovery. There is a clear need for more research into the performance of single room heat recovery units before they can be used with absolute confidence.

Mechanical heat recovery ventilation used in conjunction with a heat pump is a development of the standard MVHR units. The heat pump is to all intents and purposes the same technology as a refrigerator, consisting of a refrigerant circuit, compressor, evaporator and expansion valve. It is a device that has contrasting optimum operating characteristics to the heat exchanger. While the heat exchanger works most efficiently at high temperature differences between the two fluid streams, the heat pump is at its most efficient when operating at low temperatures. Furthermore, the heat pump has the ability to reject heat into an airstream and hence raise it to a higher temperature than the airstream from which it takes the heat. This is not to say that the heat must be rejected to air. A water circuit can also be used. Heat pumps are therefore at their best when recovering low-grade heat. They are successfully used in several commercial applications, one of the most well known being the cooling of cellars in public houses, which have the large amounts of heat.

With knowledge of the optimum deployment of heat pumps, it was a logical step to try and further increase the efficiency of domestic heat recovery by the use of heat pumps. The heat-recovery unit and heat pump are used in series, with the heat pump recovering extra heat from the exhaust airstream, which the heat exchanger cannot deal with. Extremely high recovery efficiencies can be achieved using such systems; well in excess of 85% is a realistic expectation.

There is unfortunately a downside to this apparently successful device. The rejection of heat from the heat pump is high, and it can recover large amounts of heat on a continuous basis. If it were to be used to heat the domestic hot-water supply, then far more hot water would be produced during the course of a day than a typical household would use, and furthermore the water might well be at a dangerously high temperature. Installing both a mechanical ventilation system and a wet central-heating system would be totally uneconomic, and therefore the idea of rejecting the heat to the central-heating circuit can be discounted. If the heat were rejected to a supply airstream for re-heating the house, then the supply air would enter the house at a high temperature, perhaps as high as 70°C. Such a supply temperature would be extremely hazardous to the occupants. The only way to temper the air temperature would be to mix the hot air with cold incoming air, but then over-ventilation would be taking place. Heat pumps are designed to run at an essentially constant rate. The scope for reducing heat recovery by turning down the heat pump is even less than with a heat-recovery unit. Under the wrong psychrometric conditions, water condensing on the evaporator side could freeze and stop the unit from functioning efficiently. The only practical way of reducing heat recovery at a heat pump is to use a by-pass arrangement as shown in Figure 5.14. A proportion of the air flow in the system is diverted past the heat pump. Whilst this removes the need to turn down the heat pump, the energy efficiency of the unit is severely impaired, as not all the ventilation air is passing through it.

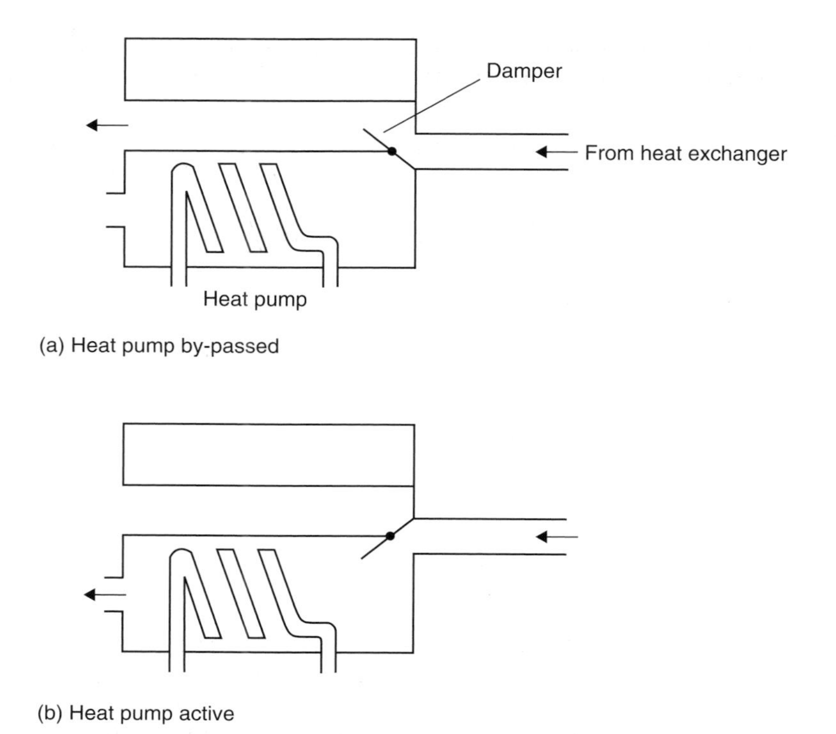

Figure 5.14 By-pass arrangement around heat pump

5.10 The calculation of true efficiency values for mechanical ventilation with heat recovery

As was briefly mentioned in Section 5.9, the heat exchanger within an MVHR system can only recover heat from that proportion of the total ventilation air passing through it. The significance of this fact is quite considerable, as the following calculation shows.

For the purposes of the calculation, the following basic design values are taken:

- *House size*: 350 m^3, in other words, a typical small terraced property. The house has three bedrooms, one kitchen, one bathroom/WC and two downstairs rooms.
- *House fabric air leakage rate*: Calculations to be performed at 3, 5, 7, 10 and 12 air changes per hour at 50 Pa applied pressure difference (Q_{50}). These values will be reduced to an air leakage rate under "typical" conditions by the well-known rule of thumb which merely divides the Q_{50} value by 20.

- *House ventilation system*: Central system designed to provide 1 air change per hour on a continuous basis, in this case 250 m^3/h. The efficiency of heat recovery is assumed to be 65%.
- *Temperatures*: An internal–external temperature difference of 15°C is assumed.

The real efficiency of an MVHR system, ε_{real}, is given by:

$$\varepsilon_{real} = \frac{H_{MVHR}}{H_{total}} \tag{5.1}$$

$$= Q_{MVHR}\varepsilon_{bench}\frac{T_i - T_o}{3} \tag{5.2}$$

$$= (Q_{MVHR} + Q_{inf})\frac{T_i - T_o}{3} \tag{5.3}$$

where H_{MVHR} is the heat recovered by the MVHR unit; ε_{bench} is the bench efficiency of the MVHR unit; H_{total} is the total ventilation heat loss and Q_{inf} is the air infiltration through the fabric (m^3/h).

Figure 5.15 shows the effect of envelope airtightness on the real efficiency of MVHR. This reduction in real efficiency might at first sight seem surprising, but it is borne out by experimental observation, for example.[35] These results imply that the real efficiency of an MVHR unit is likely to be less than two-thirds of the bench efficiency of the unit in the majority of dwellings in the UK, given that the mean Q_{50} value for the housing stock is of the order of 12 air changes per hour at 50 Pa applied pressure difference. The energy savings achieved will, therefore, be substantially less than manufacturers' efficiency data might suggest.

For the case of the single-room heat-recovery unit, the situation is more complex. If an installed unit or units were allowed to run continuously, then the efficiency calculation would be much the same as for a central unit. However, as has been mentioned previously, it would appear that the intention is to use some sort of humidity control for boosting air flows when relative humidities are high, but otherwise to allow the units to provide background ventilation. Due to this the basis for calculation is much more difficult to define.

Q_{50} (ach at 50 Pa)	3	5	7	10	12	15
ε_{real}	0.565	0.520	0.481	0.433	0.406	0.371

Figure 5.15 True efficiency of MVHR unit

The following extra data items have to be defined in order to make an estimate of the overall efficiency of single room units:

- *Background ventilation*: This will have to be provided in each of the other rooms in which single-room units are not installed. In this case, this will be in the three bedrooms and the two downstairs rooms. A typical trickle ventilator provision will give about 25 m^3/h per ventilator, resulting in this case in a total of 125 m^3/h. In the kitchen, it will be assumed that the single-room unit will give a trickle rate of 60 m^3/h at a recovery efficiency of 75%, whilst in the bathroom the smaller unit will give a trickle rate of 25 m^3/h at a recovery efficiency of 60%.
- *Boost ventilation*: The kitchen unit will be assumed to provide 225 m^3/h at a recovery efficiency of 65%, whilst the bathroom unit will be assumed to provide 75 m^3/h at a recovery efficiency of 60%.

Each unit will be assumed to run at the boost rate for 2 h per day.

The calculation is made over the daily basis. In this case, the true overall heat-recovery efficiency, $\varepsilon_{\text{true}}$, is given by:

$$\varepsilon_{\text{true}} = \text{total heat recovered during the time period/total ventilation heat loss during the time period}$$

$$= \text{[total heat recovered during boost periods + total heat recovered during trickle periods]/ total ventilation heat loss during the time period} \quad (5.4)$$

$$= [(\Phi_{\text{bkitch}}Q_{\text{bkitch}}\varepsilon_{\text{bkitch}} + \Phi_{\text{tkitch}}Q_{\text{tkitch}}\varepsilon_{\text{tkitch}} + \Phi_{\text{bbath}}Q_{\text{bbath}}\varepsilon_{\text{bbath}} + \Phi_{\text{tbath}}Q_{\text{tbath}}\varepsilon_{\text{tbath}})(T_{\text{i}} - T_{\text{o}})/3]/[(Q_{\text{inf}} + Q_{\text{trickle}} + \Phi_{\text{bkitch}}Q_{\text{bkitch}} + \Phi_{\text{tkitch}}Q_{\text{tkitch}} + \Phi_{\text{bbath}}Q_{\text{bbath}} + \Phi_{\text{tbath}}Q_{\text{tbath}})(T_{\text{i}} - T_{\text{o}})/3] \quad (5.5)$$

$$= [(\Phi_{\text{bkitch}}Q_{\text{bkitch}}\varepsilon_{\text{bkitch}} + \Phi_{\text{tkitch}}Q_{\text{tkitch}}\varepsilon_{\text{tkitch}} + \Phi_{\text{bbath}}Q_{\text{bbath}}\varepsilon_{\text{bbath}} + \Phi_{\text{tbath}}Q_{\text{tbath}}\varepsilon_{\text{tbath}})]/[(Q_{\text{inf}} + Q_{\text{trickle}} + \Phi_{\text{bkitch}}Q_{\text{bkitch}} + \Phi_{\text{tkitch}}Q_{\text{tkitch}} + \Phi_{\text{bbath}}Q_{\text{bbath}} + \Phi_{\text{tbath}}Q_{\text{tbath}})] \quad (5.6)$$

where Φ_{bkitch}, Φ_{tkitch}, Φ_{bbath} and Φ_{tbath} are the proportion of time that kitchen and bathroom system on boost and trickle settings, respectively.

Q_{bkitch}, Q_{tkitch}, Q_{bbath} and Q_{tbath} are the air infiltration of the kitchen and bathroom systems on boost and trickle settings, respectively.

$\varepsilon_{\text{bkitch}}$, $\varepsilon_{\text{tkitch}}$, $\varepsilon_{\text{bbath}}$, $\varepsilon_{\text{tbath}}$ are the heat-recovery efficiencies of the kitchen and bathroom units at the boost and trickle settings, respectively.

It should, therefore, be noted that real efficiency is independent of the internal–external temperature difference.

For the purposes of this calculation, Φ_{bkitch}, $\Phi_{\text{bbath}} = 0.167$ (2 h out of 24 total usage) and Φ_{tkitch}, $\Phi_{\text{bbath}} = 0.833$.

Q_{50} (ach at 50 Pa)	3	5	7	10	12	15
ε_{real}	0.267	0.244	0.225	0.201	0.188	0.171

Figure 5.16 True efficiency of single-room MVHR unit

Figure 5.16 shows the results of calculations of ε_{real} for the scenario given above. The results indicate that the overall efficiency of heat recovery using the two single-room units is less than half that for a central heat-recovery system serving the same dwelling to the requirements of Part F. This is due in the most part to the need for provision of trickle ventilation in other rooms. For the range of typical values of dwelling air leakage, the units are recovering less than 20% of the heat loss due to ventilation. The results are calculated on a set of suppositions that are typical but likely to some variation amongst the housing stock. As a guideline, increasing the dwelling volume will further decrease the real efficiency of heat recovery. Increasing the amount of time for which the units are operating on boost setting will increase the real efficiency, but by relatively little. At the time of writing, there is little, if any, site data for the performance of single-room heat-recovery systems in the real dwellings.

Efficiency of heat recovery is only one indicator of performance. As has been previously mentioned, the most commonly used means of expressing the overall efficiency for devices such as refrigeration plant is the COP. In the case of a refrigerator, the COP would be the heat removed divided by the power taken by the compressor. The same principle can be applied to heat-recovery devices. As when calculating recovery efficiencies, the COP of a device can be expressed either as that of the device itself, or else for the total system in which the device is operating, in this case, of the dwelling in which the device (or devices) in the case of single room units. The values for installed systems are of more interest within this context.

For the central system described above, a typical fan power consumption would be of the order of 200 W. This means that at a bench efficiency of 65% as per previously assumed, the maximum COP of the system at an internal–external temperature difference of 15°C would be given by:

$$\begin{aligned} \text{COP} &= \frac{250 \times 0.65 \times 15/3}{200} \\ &= 4.0625. \end{aligned}$$

In practice, the real COP does not vary with building air leakage rate in the same way as the real recovery efficiency, since the actual amount of heat recovered is to all intents and purposes the same. Analysis of the performance data for the single-unit scenario shows a similar pattern: in this case, the real COP is of the order of 4.3, which is slightly better than for the central system. These results underline the fact that overall COP is not a reliable indicator of the overall pattern of energy consumption, as it does not take into account the overall picture of ventilation loss. In effect, the heat-recovery devices will sit in glorious isolation recovering heat, oblivious to the heat losses resulting from fabric air leakage.

The modern trend is to analyse matters of energy consumption from the viewpoint of carbon dioxide emissions, as the issues associated with *greenhouse gases and global warming*. It is, therefore, timely to revisit previous concepts of COP, and to look at energy savings from the view of an environmental COP (ECOP). An environmental COP might be defined as:

$$\text{ECOP} = \frac{\text{CO}_2 \text{ emissions saved}}{\text{extra CO}_2 \text{ emitted}}. \qquad (5.7)$$

When ECOP is considered, the results obtained look very different. This is due to the fact that the result is influenced not only by the fuel being saved but also the fuel being expended in making the saving. In the majority of properties, gas is now the principal fuel for space heating. Even one of the less sophisticated gas boilers can operate at 70% efficiency, whilst a properly controlled condensing boiler can achieve efficiencies of over 90%. The only fuel that can be used for powering a fan is electricity. In the majority of cases, this will come off the national grid. Allowing for the mix of sources, the overall efficiency of electricity production from the national grid is about 30%. This means that the carbon dioxide emissions from electricity production are 2.3 times greater per unit than is the case for space heating from gas. Applying the principle to the energy savings achieved by a central heat-recovery unit gives an ECOP value of about 1.8 for the single-room unit scenario when using a standard boiler. If a high-performance condensing boiler is used, then the ECOP falls to as low as 1.4. It should be, of course, noted that if the means of heating is electricity, then the ECOP should be the same as the standard energy COP.

All the calculations carried out use values for efficiency of heat recovery taken from manufacturers data. As has been previously mentioned, these are based on bench tests of new equipments. It is very likely that under real conditions these performances will not be achieved. This will have the effect of driving down the ECOP even further. Coupled with other uncertainties such as the degree of ventilation efficiency, there seems to be a real danger that in certain circumstances the ECOP could fall below unity; in other words, more carbon dioxide was being emitted in running the heat exchanger fans than was being saved in the form of ventilation heat recovery.

Despite changes to the means of calculation, the National Home Energy Rating (NHER) software shows very little difference in performance between MVHR and other more simple alternatives of ventilating dwellings. At the moment, single room units are not included as part of the NHER assessment method, but it would be reasonable to suppose that the difference in rating outcome between these units and other methods would be even smaller on account of the greatly diminished overall heat recovery efficiencies.

References

1 The Building Regulations for England and Wales, Approved Document F, HMSO, 1995.
2 P Jones, P O'Sullivan. The role of trickle ventilators in domestic ventilation design. *7th AIVC Conference, Occupant Interaction with Ventilation Systems*, Supplement to Proceedings, October 1986, Stratford on Avon, UK, AIVC, pp. 91–97, 1986.
3 Energy Efficiency Office. Trickle ventilators in low energy houses. Energy Efficiency Demonstration Scheme Expanded Project Profile 109. BRECSU, October 1985.
4 M Bassett. Passive ventilators in New Zealand homes. Part 1. Numerical studies. Air Infiltration and Ventilation Centre, 1994, *The Role of Ventilation, 15th AIVC Conference*, Buxton, UK, September 1994, Vol. 1, pp. 35–56.
5 M Bassett. Passive ventilators in New Zealand homes. Part 2. Experimental trials. Air Infiltration and Ventilation Centre, 1994, *The Role of Ventilation, 15th AIVC Conference*, Buxton, UK, September 1994, Vol. 1, pp. 35–56.
6 I Ridley, J Fox, T Oreszczyn, SH Hong. The impact of replacement windows on air infiltration and indoor air quality in dwellings. *International Journal of Ventilation* 2003, 1(3), 209–218.
7 CY Shaw, A Kim. Performance of passive ventilation systems in a two-storey house. *5th AIC Conference, The Implementation and Effectiveness of Air Infiltration Standards in Buildings*, Reno, Nevada, 1–4 October 1984, pp. 11.1–11.27.
8 N Bergsoe. Investigations on air change and air quality in dwellings. CIB W67 Symposium, *Energy, Moisture and Climate in Buildings*, September 1990, Rotterdam, p. 2.3.
9 A Hardy. *Northern Housing Consortium Report*, 1983.
10 TRADA Technology Ltd. *Timber Frame Construction*, 2nd edition. TRADA, 1994, ISBN 0 901 348 94 5.
11 A Gaze, K Johnson, C Brown. A passive ventilation system under trial in UK homes. *6th AIC Conference, Ventilation Strategies and Measurement Techniques*. Het Meerdal Park, The Netherlands, September 1985. Air Infiltration Centre, 1985. pp. 4.1–4.27.
12 A Gaze. Passive ventilation: a method of controllable natural ventilation of housing. TRADA Research Report. 12/86, TRADA, 1986.
13 C Irwin, RE Edwards. Further studies of passive ventilation systems – assessment of design and performance criteria. *9th AIVC Conference*, Gent, Belgium, September 1988.
14 V Meyringer, L Trepte. Requirements for adequate and user-acceptable ventilation installations in dwellings. *7th AIVC Conference, Occupant Interaction with Ventilation Systems*, September 1986, Stratford on Avon, UK, Bracknell, AIVC, 1986, pp. 1.1–1.10.
15 C Uglow, RK Stephens. Passive stack ventilation in dwellings. *Building Research Establishment*, IP21/89, 1989.

16 L Parkins. Experimental passive stack systems for controlled natural ventilation. *CIBSE National Conference* 1991, Canterbury, April 1991, pp. 508–518.
17 A Cripps, R Hartless. Comparing predicted and measured passive stack ventilation rates. *15th AIVC Conference,* Buxton, UK, September 1994, Vol. 2, pp. 421–430.
18 J Palmer, L Parkins, P Shaw, R Watkins. Passive stack ventilation. *15th AIVC Conference,* Buxton, UK, September 1994, Vol. 2, pp. 411–420.
19 WF de Gids, HPL den Ouden. Three ivestigations of the behaviour of ducts for natural ventilation, in which an examination is made of the influence of location and height of the outlet, of the built-up nature of the surroundings and of the form of the outlet. Building Research Foundation, Netherlands, UDC 533.6.07:697.921.2:728.2/.3. (1985).
20 P Welsh. The testing and rating of terminals used on ventilation systems. *5th AIVC Conference,* Buxton, UK, September 1994, Vol. 1, pp. 371–380.
21 JEF Van Dongen. The influence of different ventilation devices on the occupants behaviour in dwellings. *11th AIVC Conference, Ventilation System Performance,* September 1990, Belgirate, Italy, Proceedings published March 1991, Vol. 2, pp. 101–120.
22 RK Stephen, LM Parkins, M Woolliscroft. Passive stack ventilation systems: design and installation.
23 RE Edwards, M Entwistle. Use of tile terminals as passive stack system terminals. DOE Report, 1995.
24 RE Edwards. The influence of Bend angles upon the performance of Passive Stack duct systems within dwellings. *International Journal of Ventilation* 1(3), 225–231.
25 *ASHRAE Fundamentals Handbook: Duct Design.* Chapter 34, American Society of Heating and Air Conditioning Engineers, 2001.
26 RE Edwards. *International Journal of Ventilation* (to be published).
27 Energy Efficiency Office. Energy-efficient ventilation in housing. *Good Practice Guide* No. 268, DETR, 1999.
28 RK Stephens, L Parkins, M Wooliscroft. *Passive Stack Systems: Design and Installation* BRE Information Paper 13.94, Building Research Establishment, 1994.
29 Building Regulations for England and Wales: Approved Document B, "Fire". HMSO, 1991.
30 J Francis, RE Edwards. The use of infrared controlled air terminals in an office building. *CIBSE/ASHRAE Joint National Conference.* Part 2, September 1996, Vol. 2, pp. 311–320.
31 Barthez M, Soupault. Control of ventilation rate in Buildings using H_2O and CO_2 content. *Proceedings of the International Seminar on Energy Saving in Buildings.* The Hague, pp. 490–4, 1983.
32 Wouters P, Vandaele L. Experimental evaluation of a hygroregulating natural ventilation system. *11th AIVC Conference, Ventilation System Performance,* September 1990, Belgirate, Italy. Proceedings published March 1990, Vol. 1, pp. 149–156.
33 C Irwin, RE Edwards. Multiple cell air movement measurements. *6th AIC Conference, Ventilation Strategies and Measurement Techniques,* The Netherlands, September 1985. Air Infiltration Centre, 1985. pp. 8.1–8.18.

34 M Woolliscroft. The relative energy use of passive stack ventilators and extract fans. *15th AIVC Conference*, Buxton, UK, September 1994, Vol. 1, pp. 245–256.
35 SL Palin, R Winstanley, DA McIntyre, RE Edwards. Energy implications of domestic ventilation strategy. *14th AIVC Conference, Energy Impact of Ventilation and Air Infiltration*, Copenhagen, Denmark, 21–23 September 1993, pp. 141–148.

6

Specialist Ventilation Strategies

6.1 Ventilation in pitched roof spaces

The majority of roofs in housing within the UK are some variation of the pitched type. Harrison[1] identifies four categories of pitched roofs:

1. The *dual-pitched* (also known as *duopitch*) roof, where the roof is a triangulated structure spanning between two exterior walls.
2. The *portal frame,* where both the walls and roofs are supported by means of a single structural member.
3. The *single-pitched* (also known as *monopitch*) roof, where a roof slope is formed by one of the two means, either by simple beam or purlin laid to slope with no tie at the lower end, or else by a single-pitch truss, which is carried on two walls.
4. The *troughed dual pitch,* in which the roof valley runs in the centre of the building, usually normal to the span of the supporting members.

The roofs of most dwellings fall into the first category. Variations of the duopitch roof include gabled and hipped types. Some properties have monopitch roofs for their entire span. Abutments, such as outhouses and extensions sometimes have a roof of the monopitched type, although it is more often the case that extensions make use of flat roofs. In many cases, the householder eventually pays dearly for the choice of the latter. One of the main advantages of the pitched roof is that it can be used for large spans without support, thereby saving on floor space below. The use of the triangular truss in whatever type affords maximum capacity for the carrying of load.

The vast majority of roof spaces has until the recent past remained unused as occupied spaces, although in some cases they are pressed into service for a variety of purposes ranging from the storage of rarely used household items (such as Christmas decorations) to the housing of large model railway layouts. Notable exceptions to this are the large number of three-storey houses with rooms in the roof found (e.g. in South Yorkshire) and the practice of using part of the space under the roof pitch for a so-called dormer bedroom within bungalows. Some householders elect to convert lofts into extra habitable rooms. This raises issues of regulatory compliance, in particular fire safety and structural loading. Some local authorities are reluctant to give planning permission for such conversions. In such properties, the volume of roof space is significantly reduced. It is most probable that growing pressure on house plan areas as a result of recent Government pronouncements will mean that increasing the use will be made of space in the loft as a location for a room. The use of the roof space as a habitable space does not mean that problems with ventilation do not exist. These will be dealt with in more detail later on in this chapter.

Roofs may have one of the two generic types of roof covering. These are permeable (e.g. roofing tiles) or else impermeable (e.g. sheet material). In the case of the latter, the passage of air into the roof space will be inhibited by the sheet material. In the case of permeable roof coverings, significant passage of air into the roof will occur as a result of wind-induced pressure differentials. With this goes a risk of fine particles of driven snow and rain being driven into the roof space in suitably adverse weather conditions. This problem can be overcome by means of the use of sarking felt.

A section of a typical pre-war-pitched roof of English construction is shown in Figure 6.1. It will be noted that sarking felt does not form part of the construction. Prior to World War II, the use of sarking felt as an underlay on pitched roofs was almost unheard of. Apart from attempts to reduce leakage around tiles by means of back pointing, roof spaces had large amounts of open area which provided high levels of ventilation. The 1991 English House Condition Survey[2] notes that about 1 in 6 of all houses in the sample group had no sarking, 1 in 2 had a felt-type sarking material, 1 in 50 had polyethylene sheeting, 1 in 500 had sprayed coatings and 1 in 50 had timber-sarking boards. The use of sarking board is the most common practice in Scottish dwellings, where weather conditions are likely to be more severe on average than in the rest of Great Britain. A major advantage of sarking boards is that unlike flexible sheet materials, they do not billow in high winds and provide a means of snow ingress.

The introduction of sarking felt as a means of eliminating the penetration of wind-driven snow served to drastically reduce the air-change rate within the roof spaces. Field measurements indicate that this reduction could be as much as 80%.[3] This implies that there is much less tendency for any accumulated water vapour within a pitched roof space to be removed by ventilation.

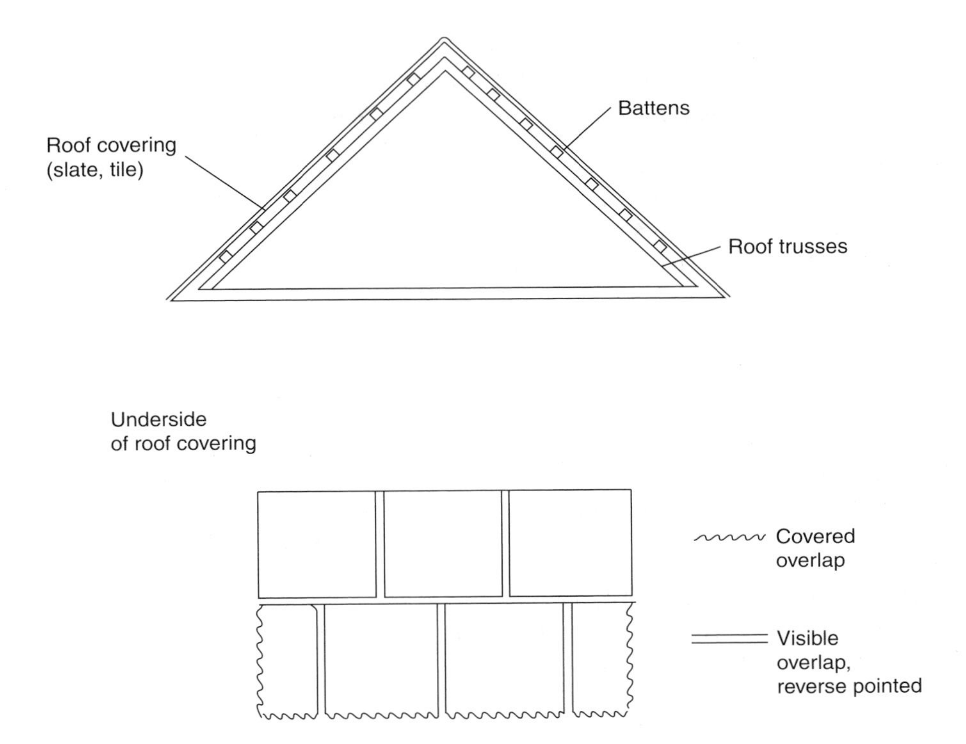

Figure 6.1 Section of pre-war English-pitched roof

The reduction in ventilation would not in itself have been a problem. However, other matters also arose. The "discovery" of energy efficiency in the early 1970s led to the widespread installation of low-level insulation in pitched roof spaces, creating what are classified as *cold-pitched roofs*. Such an arrangement is shown in Figure 6.2. At the time, this was rightly identified as an easy to install and very cost effective means of achieving significant energy savings. However, its use within a roof space would also be responsible for a decrease in structural temperatures. Coupled with the extra-moisture production in the occupied space and the reduction in roof-space ventilation rates, the effect of this was to increase the risk of condensation within the roof space itself.

It is sometimes not appreciated that what extremes of temperature can be observed within a pitched roof space. In the summer, air temperatures can quite easily exceed 45°C, whilst roof surfaces can become quite literally hot enough to fry an egg. In cases where electrical cables have been placed below insulation in direct contravention of recommended practice, overheating of the cabling to the point of being a fire hazard can occur. Summer conditions are mercifully not going to cause condensation problems within pitched roof spaces. The real difficulties arise in winter. In dull conditions, air and structural temperatures will be similar to those

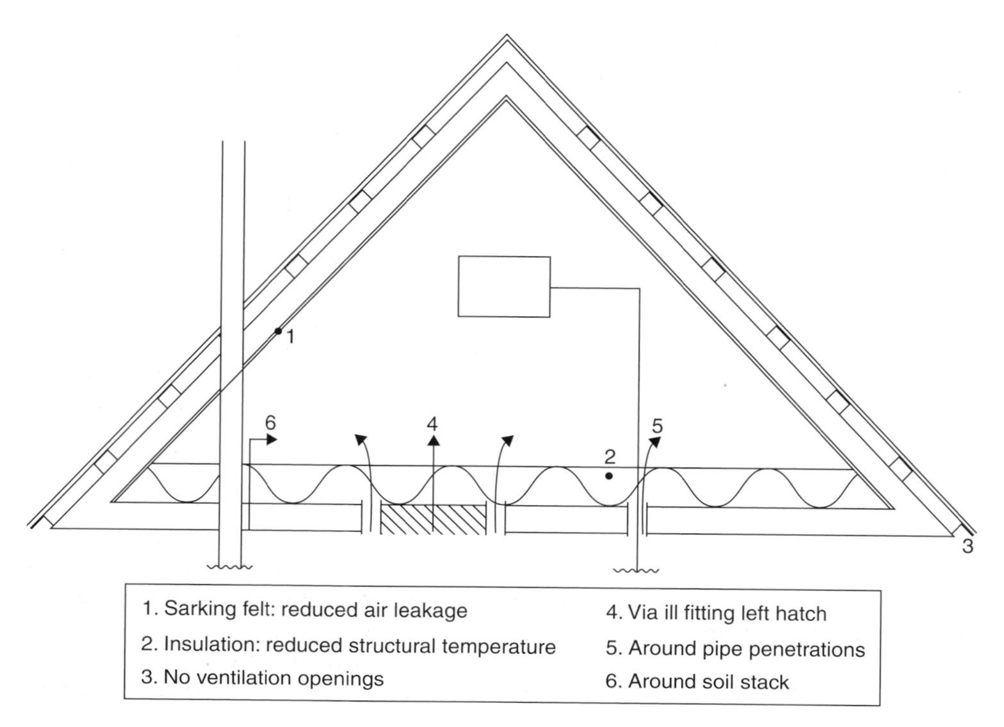

Figure 6.2 The cold-pitched roof of the early 1970s showing the factors leading to condensation risk and air leakage routes into the roof space

of the external environment. On clear nights, quite surprising temperature drops can be observed. The main mechanism for heat loss is by radiation to the clear night sky. The resulting temperatures can be very low. The underside of tiles or sarking felt can quite easily cool down to as low as −15°C. This provides a site for condensation to happen. In extremes of cold, any water condensing may freeze, and a layer of ice may build up if the cold weather is prolonged over a period of several days or more with no thawing during the day.

Moisture transfer into the roof space is a key consideration. As has been noted earlier, moisture production within the occupied space can be as high as 15 kg/h.[4] A proportion of this will find its way into the roof space in the form of water vapour. Figure 6.2 summarises the technical factors leading to increased condensation risk, and shows the main routes for water vapour ingress into the roof space. The primary means of water-vapour transfer is by air movement via cracks in the ceiling, ill-fitting loft hatches and poorly sealed service penetrations, such as water pipes and ceiling roses.

The effects of condensation may well not be apparent within the occupied space. Nearly all cold-pitched roofs will at some time experience some condensation, but ventilation will be adequate to evaporate it when condensation stops and to dry off any dampness of materials. The timbers used in the construction of the roof have a large capacity for buffering water-vapour accumulation, and therefore the effects of condensation of high air-moisture contents of roof space may be ameliorated. It should not be forgotten that prolonged exposure to high-airborne-moisture contents will cause timber to rot. It is usually the case that if the timber-moisture content rises above 18% for any periods of time, then there is a risk of wet rot. Even if the timber has been treated, the benefits of the treatment will eventually be lost under conditions of prolonged exposure to high-moisture content. The effects of timber rot are well documented (e.g. refer to reference[5]): suffice to say that mould attack and rotting may eventually lead to structural failure if undetected or ignored. Remedial work is expensive and ultimately provides no lasting solution to the condensation problem if the factors causing the problem are not dealt with as part of the remedial work.

Serious condensation problems are much more obvious. The manifestation of the problem is likely to be akin to the sound of rain within the roof space. Water drips onto the insulation, reducing its effectiveness. The effects of water entering electrical installations and equipment could be dangerous. If accumulations of frozen condensate thaw out once the weather improves, the effect will be more dramatic, with the likelihood of water penetration into the occupied space below and resultant decorative, and possibly cause structural damage to ceilings.

6.2 Control of condensation in cold-pitched roofs

It must be accepted that, short of hermetic sealing of the roof, some condensation is inevitable. The strategy should be to address the basic issues of ventilation strategy and the movement of water vapour into the roof. The means of ventilation should ensure that adequate cross-flow ventilation is provided.

Much has been made of the role of the ridge tile ventilator in eliminating stagnant pockets of air at high level[6]. Supporters of the use of ridge ventilation have led a concerted campaign to promote the use of high- and low-level ventilation. On the basis of it, the idea of the stagnant pocket of air is quite plausible, and the consequences of such a phenomenon occurring would, doubtlessly, be damaging to the building fabric. It has to be said, however, that the author has been unable to unearth a field study in which the phenomenon of the high-level stagnant pocket has actually been observed.

The most striking piece of research into roof-space condensation and the influence of ventilation strategy was carried out by the then University of Manchester Institute of Science and Technology (UMIST) on behalf of the

Building Research Establishment (BRE) in the early 1980s. The work was instigated in recognition of the fact that there was little if any understanding of the subject that could be used to assess the effectiveness of the then prevailing Building Regulations. The results, based on the most part of the tracer-gas measurements of air-change rates within the roof spaces of a sample of different properties, indicate that the use of soffit-only ventilation provides adequate levels of ventilation within roofs. Air-velocity profile measurements over a period of 2 years in a test house-forming part of a sample group of properties in an energy-monitoring study lead to the conclusion that within a duopitch roof with a pitch of less than 30°, there is no evidence of stagnant pockets of air at high level. No statistical significance could be attached to any of the measured variations in air-velocity distribution. Subsequently, it has been demonstrated[7] that on the basis of flow visualisation and air-velocity measurements made within a model roof test facility, a key factor influencing air-flow patterns within the roof space seems to be the angle of entry of air into the roof space at eaves level. Where soffit pipes were used, it was possible to demonstrate limited (but not zero) air movement at high level if the pipes entered the roof space at too shallow an angle.

The UMIST study also investigated the role of ridge tile ventilators, but not just on air-change rates within the roof space. Using multiple tracer-gas techniques as described in Chapter 4, the variation of air flows from house with differing weather conditions and roof-space ventilation strategy were also determined. The roof-space air-change measurements showed that ridge tile ventilators, when used in conjunction with low-level ventilation, increase air-change rates at low wind velocities as a consequence of greater roof-space depressurisation. Therefore, on first consideration, ridge ventilation might be supposed to provide extra ventilation at the precise time when the risk of roof-space condensation is highest.

However, this is most certainly not the full story.

The multiple tracer-gas measurements demonstrate quite clearly that the extra depressurisation of the roof space relative to the occupied space below has the effect of increasing the flow of air from the occupied space into the roof space. Bearing in mind that the occupied space is the source of water vapour that causes the risk of roof-space condensation in the first place, the consequence of increased house to roof-space air movement is, therefore, an increased rate of water-vapour transfer into the roof. At higher roof-space ventilation rates, this will not be much of a problem. Under conditions of low wind velocity, the greater rate of water-vapour transfer will lead to a greater risk of condensation.

The follow-up to the UMIST report was indecisive. On reflection[8], the authors concede that perhaps not enough of the research findings contained within the report were put into the public domain, and hence dissemination of the findings were thus limited. The research did not really take a close enough look at the direct implications of ventilation strategy on hygrothermal conditions in roof spaces. In addition, systematic measurements of spatial variations in relative humidity were not made within test roof spaces. Again on reflection, this would have been highly interesting data to acquire. In defence of the work, electronic humidity-monitoring equipment was extremely expensive and unreliable in the early 1980s. However, it must be assumed that the findings have ultimately been put to some use by the legislators. Despite very strong lobbying by vested commercial interests, a mandatory requirement for the use of ridge ventilation in combination with low-level ventilation has not been included in any of the three revisions to Part F and the Scottish Regulations introduced subsequent to the production of the UMIST report, neither was any change made to BS5250 when it was revised in 1989. At present, there seems to be a division in the ranks of the ventilation product manufacturers. Some companies are content to lay out to potential customers the options for cold-duopitched roof-space ventilation that are open to them. Some companies (presumably without a suitable product to offer) make no mention about ridge ventilation. However, there is at least one manufacturer who is issuing marketing material (both in hard copy and via a World Wide Web site) which goes beyond merely extolling the virtues of using ridge ventilation and implies that its use is a mandatory requirement. Inspection of new-built developments leads the author to the conclusion that most builders and specifiers within the highly competitive house-building sector are astute enough to be able to make an informed judgement as to what is a regulatory requirement and what is not.

Sharples[9] approaches the issue of roof-space ventilation strategy from a different perspective. Instead of basing judgements about ventilation strategy on the results of ventilation measurements in real roof spaces, Sharples describes a set of surface pressure co-efficients measurements performed on a 1:50-scale model detached house in a wind tunnel. He points out that the measurements were also intended to provide validation data for research into the determination of surface pressure co-efficients by computational fluid dynamics. The dimensions of the model corresponded to a full-size house of 7 m by 8 m plan and of 8.25 m height. The ridge line of the house ran parallel with the longer plan dimension. The roof-pitch angle was 36.5°, which was chosen as being fairly typical of houses within the UK. The house also had an eaves overhang of 0.55 m. Pressure measurements were made at a total of 15 points on the roof of the model: five along the ridge line and five under the eaves on each side. Tests were performed at four incident wind angles: 0°, 30°, 60° and 90° to the longest plan dimension.

In his discussion of the pressure co-efficient results obtained during the tests, Sharples concludes that at incident wind angles between 0° and 30°, adequate ventilation could be obtained by the use of eaves ventilation alone. At 60° incident wind angle, windward eaves are under a small positive pressure and the leeward eaves under a uniform negative pressure. Presumably this implies that cross-flow ventilation will occur. However, at this wind angle, the leading edge positions on the ridge show large negative-pressure co-efficients in comparison to those calculated for the leeward eaves. Under these conditions, Sharples suggests that air entering the roof space at eaves level on the windward side may leave the roof space at ridge level if ridge ventilators were being used, thus resulting in adequate ventilation on the windward side of the roof space only. For 90° incident wind direction, which is parallel to the ridge line, negative-pressure co-efficients are seen at both eaves and ridge level. Sharples is of the opinion that such a regime will result in an increased tendency to suck air into the roof space from the occupied space below, with consequent increased transfer of water and greater heat losses. It seems that if ridge openings were not to be provided, then this risk would be lower; indeed, eaves-to-eaves air flow might still take place, particularly if the ceiling was fairly airtight. On the basis of this final set of results, Sharples postulates that it would seem advisable to provide some form of ventilation opening in the gable ends. This would tend to drive air through the roof space in conditions where the incident wind was blowing parallel to the ridge line.

The results obtained by Sharples are significant, despite the fact that they are confined to measurements on one house model. They back up the results presented in the UMIST work, and further emphasise the fact that ventilation at ridge height is not necessary in the typical duopitched roof. The only minor disagreement between Sharples interpretation of his results and the results obtained by UMIST relate to the performance of eaves ventilators when the wind is blowing parallel to the ridge line. The roof-space ventilation measurements of UMIST indicate that wind direction does not seem to have any significant influence on the ventilation rates. This is at odds with Sharples conclusions. However, it must be remembered that Sharples based his conclusions on the results obtained for a model of a detached house.

This is a demonstrable need for more research into this issue. However, the author is doubtful about whether it would find a sponsor.

Despite the findings of these two studies, it is most important that the use of ridge tile ventilators should not be dismissed out of hand as being an unnecessary item. This is most certainly not the case. There are some instances in which the use of high-level ventilation is highly recommended. In circumstances where a house has a compartmentalised roof space, a very wide roof span, in excess of 12 m, or where the roof pitch is very high, then the use of ridge terminals will be of great value in promoting full ventilation of the roof space.

In some cases, the use of ridge tile ventilators may be difficult. Similar problems might be encountered in terms of cutting of ridge boards as might be with the use of passive stack and/or flue terminals at ridge level. There is only a limited amount of ridge board cutting that can be done.

It is appropriate to make brief mention of monopitched roofs. Low-level ventilation alone is not appropriate for such roofs. These must be ventilated at high and low level in order to promote full ventilation. This is often achieved by the use of tile ventilators of the appropriate size and open area, although a wide range of alternative products is available.

An appropriate level of roof-space ventilation is an important feature of a successful strategy for condensation control within the pitched roof space. The other key influence on condensation risk is the flow of moisture laden air into the roof space. If this can be reduced to a minimum, then the risk is drastically reduced. One way of doing this might be to install a water-vapour proof membrane at ceiling level. This would have to be below the insulation. Whilst such a measure might seem appealing in theory, the reality is that it is very difficult to install such a membrane in a building of typical British construction. The exercise might prove to be more successful in a timber framed dwelling built under very close supervision. Even if the membrane is put in properly, there is of course no guarantee that the membrane would not be perforated during its service life (for instance, during the rewiring of the property or other building work), with consequent reduction in its effectiveness. It most certainly should not be assumed that the provision of a ceiling membrane removes the need for roof space ventilation. The use of the ceiling membrane would also have the side effect of marginally increasing the water-vapour levels within the occupied space.

A sound practical strategy for the reduction of water-vapour transfer into the roof space should concern itself with identifying potential weak spots through which transfer might take place, and eliminating these by means of good detailing and finishing. The most common routes, as have been mentioned before, are ill-fitting loft hatches, light roses in ceilings and service pipe penetrations. The first of these can be dealt with by making sure that the hatch actually fits the hole properly, and that is correctly draught stripped. A fastening device or, failing this, a heavy object on the top (such as a brick) will ensure that the hatch does not move up and down during windy conditions. When putting the hatch into position, it must be remembered that the top of the hatch itself should have insulation over it, or it becomes a potential cold bridge. The insulation should, of course, be over the heavy object used to keep the hatch in place. Some excellent propriety loft hatches with catch systems and integral insulation are now available, but are rather more expensive than the old-fashioned sheet of board. The sealing of the miscellaneous pipe penetrations can easily be achieved by means of gun-applied sealants. In the case of electrical cable penetrations, the chemical compatibility of the sealant with the cable sheathing should be ascertained prior to application.

There is one potential route for water vapour into the roof space which is not at all obvious. In the case of timber-framed houses, the batten cavity behind the dry lining will often be continuous up to ceiling level, cavity stops notwithstanding. If appropriate measures are not taken to seal out the batten cavity, then it is possible for air from the house to travel right up the cavity and thence travel into the roof space. This is clearly undesirable. It is readily dealt with by careful attention to detail.

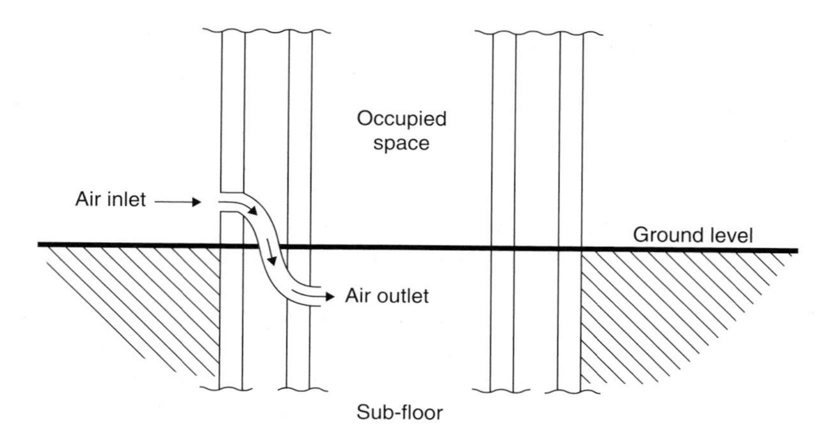

Figure 6.3 Duct flow to sub-floor space using airbricks

6.3 Sub-floor ventilation

The sub-floor cavity should in theory be built in such a way that moisture cannot accumulate. In practice, this is not the case. Provision must be made for the ventilation of the sub-floor space in order to avoid condensation-related problems and to encourage drying. This has been recognised for a long time, and ventilation of sub-floor cavities is a feature of quite old properties.

The recommendation for provision has not changed significantly for some time. Openings equivalent to $4500\,mm^2/m$ of external wall are accepted as being sufficient. The traditional airbrick in terra cotta is still available in either single or double brick size. However, in common with so many other building components, the lion's share of the modern market is taken up by plastic products supplied by a large number of firms. In situations where an airbrick cannot be located so as to allow direct passage of air into the sub-floor space (e.g. when the sub-floor space is completely below ground) the flow of air can be piped into the space by means of a ducted flow from an external airbrick inlet located above ground level, as shown in Figure 6.3.

The issue of sub-floor ventilation is also relevant with respect to the control of radon ingress.

6.4 Wall cavity ventilation

The wall cavity exists primarily to minimise the risk of driving the rain crossing from the outer leaf to the inner structure of the wall. The provision of wall cavity ventilation is an issue that is still not clearly understood.

It is also fair to say that there has been very little research into the subject. There are several sound reasons for wishing to encourage air to pass through a wall cavity, and these are not interconnected. The water content of a newly built house of traditional construction is measured in tonnes rather than kilograms. This water must be allowed to escape when it evaporates. In a dwelling of double-skin masonry construction, there will be a significant amount of water within the inner leaf. The mechanism of diffusion is not adequate to allow this water to escape. It is usual for weep holes to be provided in the external leaf to encourage drying. These used to be created by leaving vertical gaps in the mortar, but the modern tendency is to use a plastic insert. Even when dry, the masonry structure will allow the passage of water vapour from the occupied space to outside, and ventilation of the wall cavity will be helpful. Having said this, many such wall cavities have been filled with insulation without causing problems. Any difficulties that arise in such cases are usually as a consequence of water being able to bridge the cavity from the outside. It must be said that the danger of damage resulting from interstitial condensation within a masonry structure is very low.

The issue of the cavity within the timber-framed wall is altogether more complex. The hygrothermal behaviour of the timber structure is very different to the masonry wall. Large amounts of insulation are used on the inner side of the structure. Coupled with the high vapour pressure differences across the walls, this means that there is a very high risk of condensation. Almost invariably this risk will be the greatest at the interface between the insulation and the plywood sheathing. The only certain cure for this is to strive to completely eliminate the potential for water-vapour transfer into the cavity by use of an impermeable membrane on the warm side of the structure. This is the ideal case to which all timber-framed walls aspire. This being the case, ventilation of the wall cavity should have no influence on the performance of the wall. However, the wall, since it has an external leaf, still runs the risk of water passing to the internal leaf. Since the timber could become degraded by prolonged soaking, it is important to ensure that there is ventilation provision in order to encourage drying within the cavity.

6.5 Control of landfill gas ingress

The control of landfill gas is not merely an issue of controlling the indoor concentration. It should be the case that a large number of steps have been taken before the control of landfill gas within the occupied space is an issue. Development may be possible on landfill sites that have stabilised. Many landfill sites are sites of prime development potential, and there may be considerable pressure to build on such sites before the landfill has reached the stable state. Commercial buildings and blocks of flats have been built on sites which are still evolving gas. Such developments have to be undertaken in accordance with expert advice, and appropriate methane control measures must be incorporated into the design. Of these developments, difficulties have been encountered in some instances where all relevant factors were not considered at the design stage. It is recognised that it is much more likely that control measures in dwellings may be rendered ineffectual by the actions of the occupants. With the best will in the world, neither the developer nor the local authority can be expected to control developments to ensure that control measures remain effective. Furthermore, occupants may erect other structures within the confines of their land; for example, garden sheds, greenhouses, garages, summerhouses and other out-buildings. Many of these are controlled neither by planning procedures nor Building Regulations. All of these structures may act as collection points in which landfill gases may accumulate to dangerous concentrations. Owing to these difficulties, it is recognised that domestic housing should not be built on landfills that are producing significant quantities of gas. If surface measurements in excess of concentrations of 1.0% can be made, residential development should not proceed. There are difficulties associated with the translation of measured landfill gas concentrations within the ground into an assessment of the real risk to occupants of houses built on the site in question. The most important point is that high measured concentrations may not be indicative of a serious hazard if the actual gas evolution rate within the landfill site itself is low. Conversely, low measured concentrations of landfill gases within the ground may exist in association with a landfill site which exhibits a high rate of gas evolution. In summary, the concentration of landfill gases within the ground cannot be regarded as a reliable indicator of the safety or otherwise of a given site, although any such data collected will still be useful.

Assessment of the risks associated with landfill gas should not be confined to those from the ground below. The possible impingement of any landfill site within a reasonable distance of a potential development site should be taken into account. Waste Management Paper 27[10] recommends that if a development is proposed on a site within 250 m of a landfill site, whether that landfill site is operational or not, the developer must take into account the proximity of the development site to the landfill. This process must include assessments of the topography and geology of the area. The figure of 250 m should not be interpreted as a meaning that no risk could exist on a site further than this distance away. A full-site investigation should be carried out if there is a suspicion that there may be a risk of landfill gas permeation. This recommendation is also incorporated into planning Circular 17/89.

Another piece of legislation also reflecting this approach is The Town and Country Planning Act, General Development Order 1988, which sets a 30-year time threshold. The Environmental Protection Act forces local authorities to maintain registers of sites which may be contaminated.

Attention must be paid to planning permission applications involving changes of land use in the proximity of landfills.

There are a wide range of geological factors which may have an influence. Certain geological formations will encourage the lateral spread of landfill gas for quite appreciable distances from the landfill site. This is a particular problem for adjacent sites. The spread of gas or for that matter the emission rate from a site can be controlled by means of controlled pumping of the gas. In some cases, sufficient amounts of methane are generated so as for the landfill to represent a viable source of fuel. Most of the time it is the practice to burn off the gas as a flare. Great care must be taken not to pump gas from the landfill at a rate greater than the rate of gas evolution. If this happens, air will be drawn into the landfill, and the oxygen within this air will cause aerobic decomposition to recommence. This process is exothermic, and the danger is that sufficient heat may be generated to start an underground fire. A possible source of ignition for an underground fire would be a garden bonfire gas. Waste Management Paper 27[10] recommends that no residential development with private gardens should take place near a landfill site where the concentrations of methane and carbon dioxide (CO_2) exceed 1% and 1.5% by volume, respectively. This restriction exists in order to ensure that landfill gases cannot accumulate within garden sheds, greenhouses or domestic extensions, where precautions against the ingress of the gases are almost certain to be non-existent. The scope for controlling the construction of garden sheds and greenhouses following the construction of the houses themselves would be rather limited. Furthermore, it is recommended that no garden should extend to within 10 m of the boundary of a landfill site that is known to be generating gas. There is a further restriction in Waste Management Paper 27. It is also recommended that no housing development should take place within 50 m of a landfill site that is emitting landfill gas, and that gardens should not extend to within 10 m of any landfill site. In terms of measures to control risks, Approved Document C of the Building Regulations for England and Wales[11] states that if measured concentrations of methane within the ground are not likely to exceed 1% v/v, and if a house or similar small building is built with a floor constructed in the manner recommended with BRE report BRE414,[12] then no further protection measures would be required. If the measured concentration is greater than this value, then BRE414 recommends that specialist advice should be sought regarding suitable protective measures. One might suspect that the advice might in some cases be not to build on the site.

It might be tempting to assume that the methane is the only landfill gas component which is worthy of consideration in terms of control. However, this is not the case. Approved Document C2 also states that if CO_2 concentrations in the ground are in excess of 1.5% v/v, then consideration should be given to the use of the floor constructions given in BRE414. However, if the CO_2 concentration in the ground exceeds 5% v/v, then use of the recommended floor constructions is mandatory.

To achieve the objective of total exclusion in dwellings, a two-part control strategy is envisaged. This involves the removal of landfill gas from beneath the house under consideration, combined with a floor construction designed to prevent the ingress of gas into the occupied space above. The sub-floor of the dwelling should be filled with granular material. The minimum thickness of this granular layer should be 200 mm. The material used should have no fine content that might otherwise affect its permeability, and must not be heavily compacted once laid. The floor itself should be of the reinforced concrete type. Timber floors are wholly inadequate for preventing the ingress of landfill gas. The floor must also be fitted with a low-permeability membrane. Within this context, "impermeable" should be taken as with a meaning "as impermeable as possible". It is recommended that the membrane is made of polyethylene sheet, with a minimum thickness of at least 300 μm, corresponding to 1200 G. Other materials may be suitable; for example, flexible sheet roofing materials, pre-fabricated gas-impermeable barriers with welded joints and self-adhesive sheet products with a bitumen coating. The important consideration is not the absolute permeability, but rather how resistant the material is to the rigours of the construction process, as any damage thus incurred will reduce the effectiveness of the barrier. Care must be taken to ensure that joints between separate sheets of material are gastight. Particular care should be taken if sheets are welded together, as there is a danger that the membrane itself might be damaged during the process. BRE414 also points out that with careful design and selection of materials, it is possible to install a membrane which will not only keep landfill gas out, but also will act as a damp-proofing layer. Detailing is very important. The membrane should be continued across cavity walls. Cavity trays may be used to achieve this. If this is not done, then landfill gas may accumulate within the wall cavity, and in some cases may even find its way into the roof space. It is also recommended that the cavity be ventilated, although no specific advice is given as to the level of provision of ventilation openings. Particular care must be taken with service penetrations through the membrane. Ideally these should be kept to a minimum, and it is best if incoming services to the dwelling are brought into the dwellings above the floor slab through an entry with suitable sealing arrangements. In cases where it is unavoidable that services come through the floor, as it is often the case with mains' water supply pipes and soil pipes *en route* to the foul drain, then such penetrations should be surrounded in dense concrete, and a puddle flange fitted in the floor slab. The alternative is to use flexible sealant in the gap between the pipe and the floor slab. In both the cases, the membrane should be completely sealed around the service pipe as it comes up through the pipe. This is often achieved by using the so-called "top hat" section welded to the membrane. The membrane can be fitted in one of the two locations: either between the top of the granular layer and the underside of the concrete floor slab, in which case the top of the granular layer will have to be blinded in order to prevent perforation of

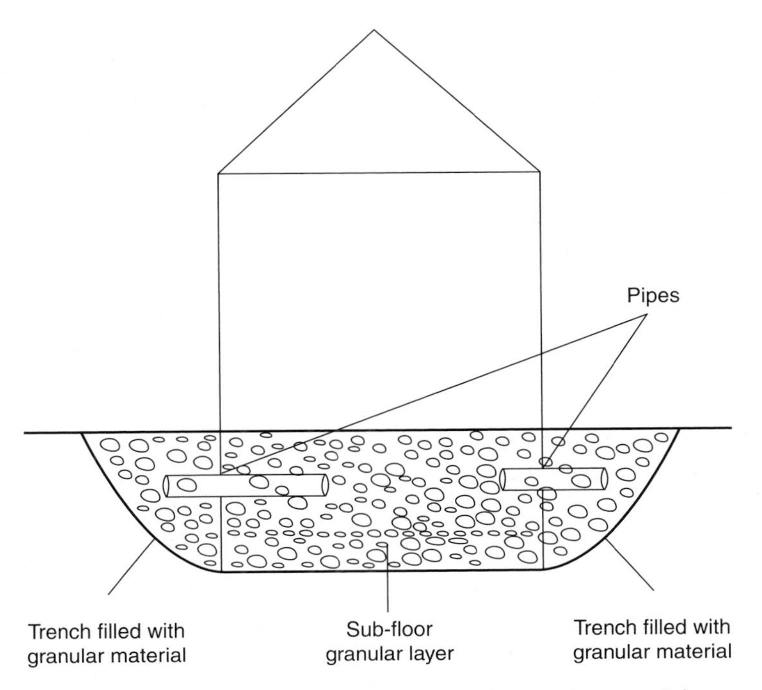

Figure 6.4 Removal of landfill gas via trench filled with granular material

the membrane; or else over the top of the floor slab, in which case a floor screed will have to be laid in order to protect it. The gas should be piped away into a surrounding trench filled with a similar granular material, as shown in Figure 6.4. Pipes should be provided at a rate of about one per metre of external perimeter wall. BRE414 is not exactly prescriptive in this respect. The gas will then diffuse away and be diluted by outside air. In practice, this solution is not perfect. Gas may be transported into the occupied space by ventilation air. The surface of the filled trench is extremely susceptible to blockage and obstruction. If the pressure of the landfill gas is significantly greater than atmospheric pressure, then the trench arrangement will not provide an adequate means of removing the gas. The concentrations at low level in the vicinity of the trench will be such that there will be a danger of re-entrainment into the dwelling. In such cases, it will be necessary to cover the surface of the trench, and to pipe the landfill gas to a discharge point above eaves level, where there is a far better prospect of the gas being swept away. BRE414 states that a suitable outlet device should be used in order to encourage the flow of gas. The example given of such a terminal is a rotating cowl type. Other variations on the basic concept of the granular layer are also permitted in order to improve the extraction of gas. Use may be made of a gas-collection pipe running through the granulated layer just underneath the floor slab. This pipe would be depressurised relative to the granular layer, and therefore the landfill gas within the granular layer would tend to migrate towards it. Such a collection pipe should be of a minimum 100-mm diameter, and should either be perforated or slotted. It may be connected to the vertical extract riser itself.

Rather than using the trench/riser method, airbricks may be used to exhaust the landfill gas to the outside. If this solution is used, then great care must be taken to ensure that landfill gas cannot leak into the wall cavity of the house above. This necessitates the use of a gas-tight seal. If a telescopic airbrick is used, then care should be taken to ensure that the two halves of the device are taped together properly in order to avoid leakage. Of course, the airbricks discharge at low level. An interesting point to note is that whereas the uncovered trench discharging at low level should not be used at high landfill gas pressures, no recommendation is made for the restriction of the use of low level exhausting from the airbricks. This could possibly be a flaw in the recommendations given in BRE414.

Instead of using the granulated layer method, a sub-floor void can be used. The floor should ideally be of pre-cast concrete with a gas-proof membrane. Ventilation is by a passive regime. The recommended minimum provision of ventilation is either 1500 mm^2 of opening per metre run of wall, or else 500 mm^2/m^2 of floor area, depending on which gives the greatest area of openings. It is not stated whether the first requirement refers to the total run of wall or merely the total run of external wall, but it would be reasonable to assume that the latter is applicable, as this would be consistent with the calculation criterion used for the case of radon gas.

The usual means of providing ventilation openings would be by means of airbricks. These should be arranged so as to give good cross-flow ventilation. Air gaps should be provided in sleeper walls so as to eliminate stagnant pockets where landfill gas could accumulate, although no recommendations for the provision of such openings are given. Care must be taken to make sure that airbricks do not become obstructed, either by additional construction or by accumulating debris. BRE414 makes reference to the importance of maximising rates of ventilation. With regard to this need, brief mention is made of the need to take into account of the knowledge of local weather conditions and sheltering, and to the need to provide adequate thermal insulation of the floor. No explanation of this latter requirement is provided, but reference is made to the BRE Document titled "Thermal Insulation – Avoiding Risks".[13]

The options for removal of landfill gases from beneath the dwellings might seem at first sight to be much the same as that for radon control, but there are several important differences. For radon control, the objective is to keep indoor radon concentrations to below the 200 Bq/m^3 action level, whereas with landfill gas, the objective is to achieve total exclusion from the occupied space. Radon gas is removed from a sump, and the use of a complete granular layer is not absolutely essential. Indeed, the sump does not even have to be underneath the dwelling if this is not feasible, as would be the case in some refurbishments. Most importantly, in the case of radon, the use of a fan on the extract system is compulsory. With landfill control, the use of a fan within a system serving a dwelling is expressly forbidden. BRE414 emphasises the need for landfill protection schemes to be completely passive; that is, they should need neither mechanical ventilation nor monitoring of landfill gas concentrations. If the rate of gas emission is such that the extract rates afforded by the use an extract fan would be needed in order to give adequate control, then the preferred and safer option would simply be not to build on such a site. This is not on the grounds of ignition risk from the fan, as electrically safe fans are readily available, but rather from the recognition that there is no guarantee that the fan will continue to operate for as long as it is required to do so. It may not be maintained properly, or indeed it might simply be disconnected. Whatever the reason for the fan becoming inoperable, dangerous build-ups of landfill gas could occur very quickly. Similarly, gas-monitoring systems cannot be relied on to detect fan failure as they themselves have a requirement for maintenance in order to be absolutely certain that they are working satisfactorily. It should be noted in passing that any restrictions to the use of mechanical extraction and gas-monitoring systems do not apply to gas control installations in commercial buildings on the grounds that maintenance schedules should in principle be quite rigorous.

6.6 Control of radon gas ingress

The strategy of the control of radon might on first consideration seem to be almost the same as for the control of landfill gas. However, there is a very important difference between the two strategies. Any leakage of landfill gas into a property is a potential source of methane accumulation and therefore could result in an explosion. Landfill gas measures seek to achieve total exclusion of permeating gas from the occupied space. The situation is totally different with radon. The gas occurs naturally, and the so-called "action level" is set at 200 Bq/m^3. This means that total exclusion of radon is neither possible nor necessary. Measures therefore centre around control.

Radon control measures can be put into two groups, namely *primary* and *secondary measures*. Primary measures seek to remove radon from beneath the floor of the dwelling. Owing to the increasing number of locations in which radon emissions are being determined as being a potential health hazard, there will be an increasing requirement for remedial treatments of existing properties. The most popular measure in such cases is the radon sump. In its simplest form, the radon sump is a pit in the ground beneath the property, from which radon is extracted by means of a fan via an extract duct and discharged to the outside. A cheap means of forming the pit is by using a 600 × 600 mm paving slab and bricks, as shown in Figure 6.5. Modern plastic products are also available

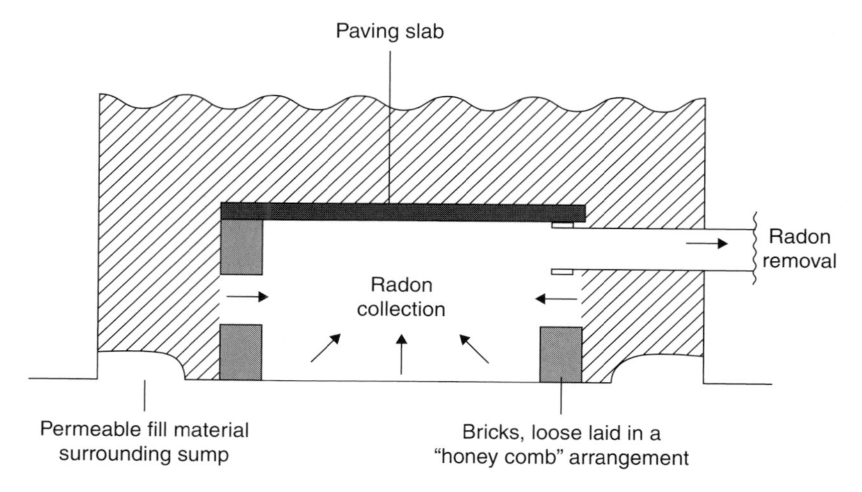

Figure 6.5 Simple radon sump

from a number of manufacturers. The duct used for extracting the radon is invariably a 110-mm diameter drainpipe, a readily available item. If used within radon systems, care must be taken to seal joints. Also care must be taken to ensure that any condensation within the ductwork could not run back onto the fan motor, as this will greatly shorten its life.

There is some latitude with respect to the location of the sump and the route by which the duct removes the radon. Figure 6.5 shows what is the preferred configuration. The sump is located as centrally as possible underneath the floor. The duct is led sideways out of the ground, and is led vertically up to the side of the house. It should be noted that the system should discharge above eaves level. This is so as to ensure that any radon leaving the system is not re-entrained into the occupied space, thereby increasing the radon levels. It may not always be the case that the sump can be located centrally. In such circumstances, it might be possible to locate the sump adjacent to an external wall, as shown in Figure 6.6. In extreme circumstances, for example where excavation might undermine the structure or foundations, another alternative exists. The sump can be sited adjacent to the building as shown in Figure 6.7. The suction from the sump will still have the effect of drawing permeating radon towards it, although its efficiency will not be as great as in the layout shown in Figure 6.6. It is possible to run the duct through the occupied space and up to high level as shown in Figure 6.8. There are likely to be problems in routing the duct unobtrusively through the house. Great care must be taken in ensuring that joints in the duct are properly sealed. It must be said that routing the duct through the occupied space should be discouraged wherever possible.

The effectiveness of primary means of protection can be improved in a newly built property. The percolation of radon towards the radon sump

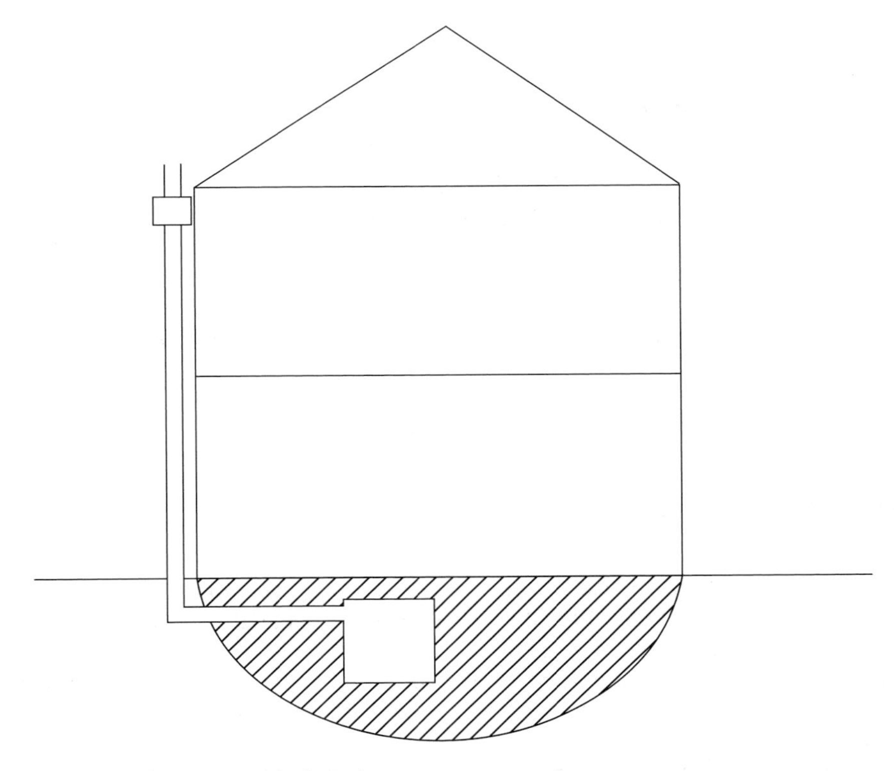

Figure 6.6 Radon sump: ideal discharge arrangement

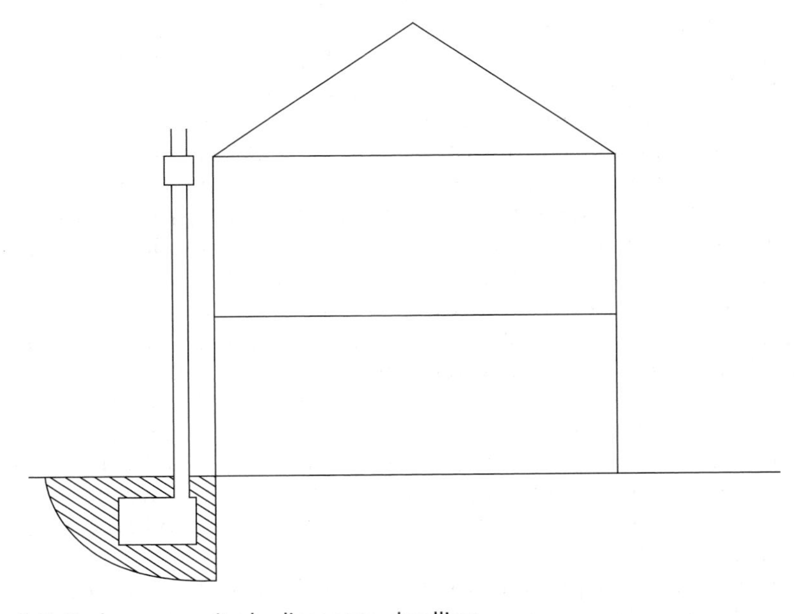

Figure 6.7 Radon sump sited adjacent to dwelling

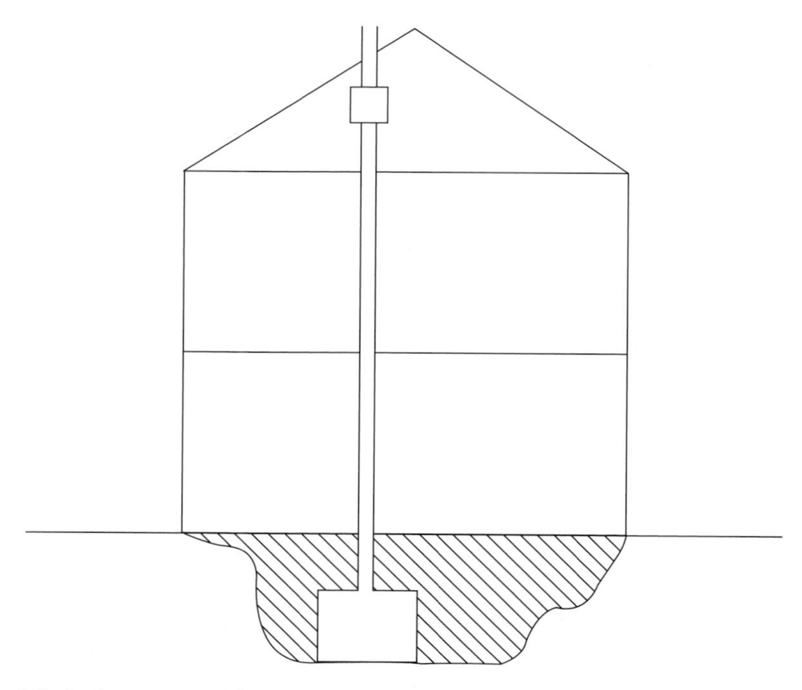

Figure 6.8 Radon sump with discharge running through the dwelling

can be enhanced by the use of a granular layer, for example of aggregate. In passing, it should be noted that extraction from radon sumps must always be provided by mechanical means. A passive stack arrangement would not be enough to give adequate radon extraction due to the high resistances to air flow involved.

Another possibility for primary protection is by ventilation of the sub-floor cavity. Cross-flow ventilation is induced in order to remove any radon percolating into the cavity. The provision of ventilation openings is set at 1500 mm^2/m of external wall perimeter, or 500 mm^2/m^2 of plan area, which ever is the greatest. The provision of openings will usually be by means of airbricks.

Secondary measures are essentially designed to prevent the passage of radon into the occupied space. This in effect means the use of membranes within the floor of the building. The membrane may be laid over the concrete floor slab, and a screed is laid over the top of it. Alternatively, the membrane may be laid on a bed of sand spread over the granular layer below the floor (so as to avoid the risk of membrane puncture) and the concrete floor poured over the top. Great care must be taken to ensure that the membrane seals out the whole cross section of the dwelling. In fact, this means that the membrane must over hang the external leaf. Of course, it is the case that any penetration of the membrane subsequent to construction runs the risk of permitting the ingress of radon into the dwelling. Other than warning the initial occupants of the danger, it is not clear how this risk could be avoided.

Choice of measures depends on the perceived risk. This is laid out in the BRE Publication BRE212 "*Radon: Guidance on Protective Measures for New Dwellings*".[14] The document sets out known areas of risk from high levels of radon emission. The areas of risk are divided into two. In the first list, the use of primary measures is deemed to be sufficient to control indoor radon concentrations to below the action level of 200 Bq/m^3. In the second list, the rates of radon emission are so great that the use of both primary and secondary measures is required.

6.7 Differences between radon and landfill gas control measures

There are more than a few equipment suppliers whose catalogues allude to "landfill and radon control product ranges", thus implying to the reader that the same kit will do both the jobs. The measures required in each case have their similarities, but in reality there are several critical differences. The difference in control strategies has already been mentioned. The most important point is that mechanical extraction of landfill gas from domestic premises is not permitted. The reason for this is that if a fan fails, then the build-up of methane gas could rapidly result in the presence of an explosive concentration. Therefore, a risk could occur in a relatively short period of time. This is not the case with radon. A failure of a fan may possibly result in an increase in the indoor radon concentration. However, even if the concentration exceeds the action level, the increase in risk of lung cancer will be relatively small, and so it is the general view that the use of mechanical fans in radon extraction systems is perfectly in order.

6.8 Flats

The definition of a flat could well be taken to be a housing unit within a building divided up into more than one housing unit. This is commonly referred to as *multiple occupancy*. In principle, any building could be used for multiple occupancy, and this includes house types which are more commonly associated with single occupancy. Examples of these include large houses that are divided up into flats and the so-called "granny flat" within a house.

There are several aspects of overall building design which must be taken into account when using buildings for multiple occupancy, such as the provision of means of escape in the case of fire. These are beyond the remit of this book. On the basis of things, the provision of ventilation is fairly a straightforward matter. It would merely be necessary to meet the requirements of Approved Document F in order to achieve Building Regulations compliance, as described in Chapter 7. In the majority of cases, this would merely mean that the provisions for a dwelling should be implemented for each flat within the building in question, and that where necessary, appropriate treatment be given to common areas.

It is fair to say that meeting the requirements of Part F will, in the case of most low-rise flats, be satisfactory. Any problems that occur will be little different to those found in single-occupancy properties. However, many multiple properties are in the rented sectors, and it is in these that problems, such as damp, condensation and inadequate heating may well be encountered in what could be said to be with disproportionate frequency. However, it is clearly the case that most of the difficulties are due to problems of affordable heat and inadequate internal temperatures, and that there are no specific deficiencies within the Building Regulations with respect to low-rise multiple-occupancy dwellings.

When taller blocks of flats are considered, matters are not so straightforward. Indeed, a definition of a "tall" block of flats is in itself far from clear. For purposes of fire safety, "tall" means three storeys or more: for buildings of this height or greater, most of the statutory requirements for the prevention of spread of flame, smoke and the provision of means of escape in the case of fire will apply. As will be seen later in this chapter, this baseline of three storeys is a very appropriate one for this particular discussion of ventilation requirements within the blocks of flats. Interestingly, three storeys is the cut-off point; beyond which those who are trying to predict ventilation rates are cautioned against assuming mean pressure co-efficients for individual building surfaces.

The block of flats is quite a common sight within Europe. In some countries, as many as 70% of the population may live within such buildings. This is entirely consistent with the tendency of many Europeans to live in smaller housing units closer to the centre of cities. In the case of the old Eastern Europe, the provision of mass housing ties in with the concepts of centralised control prevailing in such countries until relatively recently. Such housing is not without its problems, particularly within the former Soviet Union. However, the overall picture is quite positive in comparison to the UK.

Most high-rise accommodation within the UK originated in the 1960s and early 1970s. It was built as part of the overall strategy of expansion and improvement of the housing stock. High-rise, high-density housing was seen as a sensible way of providing large numbers of housing units within the urban areas. The emerging building technologies, such as precast concrete panels, were an ideal way of building high-rise flats quickly and cheaply. The standards to which the flats were built were viewed as a dramatic improvement on the conditions within the so-called slum dwellings that they were intended to replace. Of course, not all of the new housing was high rise. Many of the new estates consisted of mixtures of high- and low-rise flats.

The high-rise flats were built with the best of intentions. It would have been difficult to have foreseen the range and complexity of problems which would arise at both the social and technical levels. Many of the new estates eventually became places of high unemployment and crime. Difficulties with crime were made worse in some cases by design and layout features within the new estates. A good example of this was the deck access system within the now-demolished estates in the Hulme and Moss Side areas of Manchester. The conversion from traditional housing to flats destroyed much of the previous sense of community and social cohesion suffered as a consequence.

The difficulties encountered were not merely confined to the social category. There were a number of unforeseen technical problems that served to adversely affect the living conditions within the new flats. The first and the foremost serious problem was the actual choice of construction method itself. The use of pre-cast concrete was not confined to flats. More traditional housing forms were constructed using these systems, and also share in the technical problems. However, this fact is often overlooked (for a very good overview of the use of non-traditional-building systems in the UK, reference[15] is highly recommended). These systems made use of the new high-alumina cements, which were selected on the grounds of their excellent rapid curing properties. However, there were drawbacks to the use of high-alumina cements. The most serious drawback was the tendency of some of the chemical constituents of these cements to corrode the reinforcing bars within the panels. This resulted not only in reductions in the mechanical strength, but also the expansion resulting from the formation of iron oxide (corrosion products) caused cracking and spalling of the concrete. If this has happened, the consequences could be extremely dangerous. Setting aside corrosion problems, there were also issues of structural strength. Whereas the structures themselves were intrinsically strong from the viewpoint of traditional considerations, their resistance to blast damage was very poor. The most spectacular manifestation was the Ronan Point incident of May 1968 in which a gas explosion in a flat had the totally unexpected effect of causing the collapse of one corner of the block.

Panel crackage may have an adverse effect upon the indoor environment. A crack can be a source of moisture ingress, particularly under conditions of driving rain. The joints between panels can also be the weak points, as the materials available for sealing them at the time of construction were inadequate for the purpose and also were of poor durability. Panel-built flats are susceptible to damp problems. Given the nature of the failure mechanisms leading to the damp problem, it is often hard to effect a cure. In addition to moisture ingress, air can also infiltrate and exfiltrate the internal envelope, giving rise to excessive heat losses and hence compounding difficulties with inadequate internal temperatures. Joints are also the source of leakage problems, as the joint filling materials available at the time of construction were of poor durability. Seals within the structure are also suspected. These difficulties were often made even worse by the installation of inappropriate and/or undersized heating systems, particularly bad examples of which are under-floor electrical heating (far too expensive to run and at variance with commonly understood concepts of timer control of heating) and warm air systems (often inadequate and also requiring high levels of maintenance in order to function). During the period in which these buildings were constructed, regulatory coverage of ventilation requirements was very poor. The consequence of all these difficulties was that the majority of high-rise flats, and high proportion of system-built low-rise flats, were cold, damp and poorly ventilated.

Work of remedial measures for system-built dwellings has been ongoing for some length of time. Whereas structural integrity is a vital issue, attention has been turned in more recent years to energy-efficient refurbishment. It is difficult (and in many cases practically impossible) to apply cavity insulation in the same way as with traditional masonry structures. Indeed, even if this could be done, the resulting decrease in structural temperatures in the external structure, in this case the concrete, will increase the risk of water freezing within the concrete and hence further compromise the structural integrity. The favoured means of improving the thermal performance of panel buildings has been by means of external insulation panels. External insulation has a number of proven technical advantages over other methods of insulation. The only two drawbacks are the resistance to mechanical impacts and the durability of the materials used.

Whilst the thermal aspects of refurbishment are well refined, the same cannot be said for the ventilation aspects. The ventilation air in a flat will come from air-infiltration and ventilation systems in the same way as any other building. Panel buildings will clearly have higher air-infiltration rates due to the range of problems described above. However, little work has been done within the UK to quantify the magnitude of fabric air leakage in panel dwellings. The use of external insulation should reduce air leakage providing that joints are sealed properly. There is no codification of the need for control of air leakage beyond the good practice recommendations given in reference[16]. Again, little work has been carried out to ascertain exactly what improvements are achieved. Beyond the issue of air infiltration, the choice of ventilation system must also be considered. At present, they will go no further than choosing a solution which is deemed to comply from the options covered by the Approved Document F 1995.

Building Regulations do not address the issue of whether all its recommendations are necessarily appropriate to flats at any height within a multi-storey block. It is in reality most unlikely that this will be the case. The first and most important issue is that of the influence of building height on air infiltration. It will be recalled from Chapter 3 that wind speed varies according to height above ground level as given by Equation (3.16). This means that wind pressure increases with increasing height, but not linearly. For typical conditions, and assuming that Equation (3.16) can be reliably extrapolated to 20 m height, at a modest wind velocity of 4 m/s, the induced wind pressure at 20 m height could be greater than 20% higher than at 7.5 m height, which is more representative of a two-storey property. For taller buildings, the problems will be massively magnified. The implications are twofold:

1. Even a fairly low background fabric air leakage rate can result in significant air infiltration at elevated heights.
2. This makes the need for the provision of ventilation openings questionable at these heights.

Thus it is apparent that there is every chance that excessive ventilation will occur in flats at high levels. The problem would undoubtedly be made worse by window opening on the part of tenants, although it may be the case that this practice would be inhibited by the sheer force of the draughts that would be induced at high level by so doing.

There are further factors that complicate the issue of ventilation provision within flats. The form itself is not simple. The height of the building gives rise to significant potential for stack-induced pressure differences. Leaky risers and lift shafts may provide a route for stack-driven flows to take place. In practice, internal partitioning and fire compartmentalisation will offset the effects of this potential. In some tall buildings, pressure-relief dampers have to be provided so as to protect against possible structural damage under heavy gusts. When actuated, these dampers will affect the patterns of air flow within the building. Other buildings in the vicinity, particularly those of a comparable or greater height, will afford as degree of shelter and will thus change local pressure distribution. In the worst case, turbulent pockets of air flow will occur under certain combinations of wind velocity and direction. Such effects will give rise to unexpected fluctuations in surface pressures.

There are clearly a number of areas of uncertainty when deciding on a suitable ventilation strategy for a flat within a high-rise block. In theory, this would be an ideal situation to model. In practice, there are so many uncertainties in the actual basic data items available, surface pressure co-efficients being the most obvious example. The total date set needed for such a calculation would be extremely complex, there is no reliable procedure for the calculation of air-change rates in tall buildings which is publicly available.

6.8.1 *Practical solutions*

Notwithstanding the difficulties relating to the provision of ventilation in flats, practical solutions need to be found, and at the very least, any provision must be demonstrated to give compliance with the requirements of

Part F of the Building Regulations. It is more than likely that in most cases, compliance will be seen as the primary objective. One immediate problem relates to whether the flats in question are to be newly built or else be subjected to refurbishment. With new-built properties, the scope for designing in ventilation systems is, of course, much greater than is the case with refurbishment. Refurbishment has some potentially significant challenges associated with it. Space may well be limited, and the layout of new ventilation systems may be impeded by this. In some buildings, the structure itself may present an obstacle to the installation of systems. In certain types of construction, concrete floors may be 300 mm thick, and this will make the routing of ductwork a very labour-intensive and time-consuming process.

The first element of solution would be to provide background ventilation by means of trickle ventilators or similar openings. This is relatively a straightforward matter. Rapid ventilation could be achieved by window opening. Window openings would need to be designed with safety in mind, not only from the point of view of preventing falls out of open windows, but also to avoid the possibility of windows smashing as a result of the wind slamming them shut.

With respect to local extract ventilation in moisture-producing areas, things become more problematic. The first recourse would probably be the same as for an ordinary house, namely wall-mounted extraction fans. As most rooms in a flat will have at least one external wall, installation should in itself present in difficulties. In cases where an affected room is internal, the extract could be ducted through the outside if necessary. There are dangers associated with the use of mechanical extract fans which are an increased risk in flats as opposed to low-rise dwelling. The wind pressure on the fan discharge may be at a sufficiently high-incident wind velocity so as to stall a fan whilst it is in operation, which may lead to the fan motor burning out. If the external louvres on the fan extract are not sufficiently resistant to incident wind pressure, rain may be driven through onto the fan motor, and this may again result in failure of the motor. Another consequence of inadequate louvres may be that the fan casing becomes the source of an unwanted draught when it is not switched on, and will thus cause excessive ventilation during periods of non-operation.

Passive extract systems may be used in flats. The biggest issue concerns the layout of duct system that should be used. Unless the block-containing flats is relatively low, it will be impossible to individually terminate each passive stack ventilation (PSV) duct at roof level in the manner prescribed in BRE Information Paper IP13/94 (refer to Chapter 7). The preferred solution to this problem is not identified by UIP13/94. Recourse would have to be made to the demonstration of satisfactory performance compliance by alternative means. In this case, this would entail convincing Building Control that a particular design solution would give adequate extraction in all rooms in which it was needed.

The logical answer would be to use a system of ductwork similar to that which would be used for a mechanical extract system in a commercial building. Ducts from flats would be led into ducts of larger diameter, until one or more large diameter ducts discharge the moist air at high level. Each of the sections of the ducted extraction system will have to be sized in accordance with established procedures. The major unknowns will be the exact design values for extraction rates from each flat. The safest procedure would be to estimate values based on upper end of the range of airflow rates that might be expected. These can be produced relatively easily. In practice, the real levels of extraction will vary considerably, being affected by temperatures in the zones of extraction, the opening of windows, trickle ventilator, and internal doors and the wind pressure.

Some difficulties may arise in complex ducted PSV extraction systems. Given the relatively low-pressure differentials within such systems, there is a small possibility that a circumstance might arise where extract air travels from one ventilated room to another via the common extraction duct. Another possibility is that air from the main extraction duct might be sucked back into a ventilated room via an extraction terminal. In such complex systems, it is very difficult to predict with absolute certainty the circumstances in which either of the above-mentioned sets of circumstances might be encountered. The risk of air transfer from room to room can be minimised by ensuring that branch duct entries into a larger duct are staggered. Reverse flow cannot really be designed out, rather than by recourse to mechanical devices which inhibit the flow of air in the wrong direction. Given the low pressures in the PSV systems, it may in practice be difficult to make such devices work properly. It is more than likely that flow reversal will be transient in nature, and will not lead to any risks of condensation in the long run. In Europe, some use has been made of PSV systems which make use of the so-called "shunt" arrangement[17], as shown in Figure 6.9. Instead of being directly attached to the main extract duct, the PSV system discharges into an intermediate space formed in the main duct. This decoupling seems to act as a buffer against pressure fluctuations and undesirable flow effects.

The energy efficiency of PSV systems can be improved in exactly the same way as for low-rise dwellings in the form of humidity-modulated extract terminals. Wouters *et al.*[18] report on a Pan-European-monitoring exercise, which seems to suggest energy savings of up to 25% when humidity-modulated terminals are used. Interestingly, the results presented in the report show a decrease in energy efficiency with increasing height in the buildings monitored, serving to emphasise the points made previously about the effect of building height on system performance.

If in a particular case it is felt that reliance should not be made on passive ventilation mechanisms, then an extraction-only system can be set up, using a similar duct arrangement to the passive system, but making use of a fan to draw air out of the building. This has the advantage of guaranteeing a known extraction rate, providing that the system is commissioned

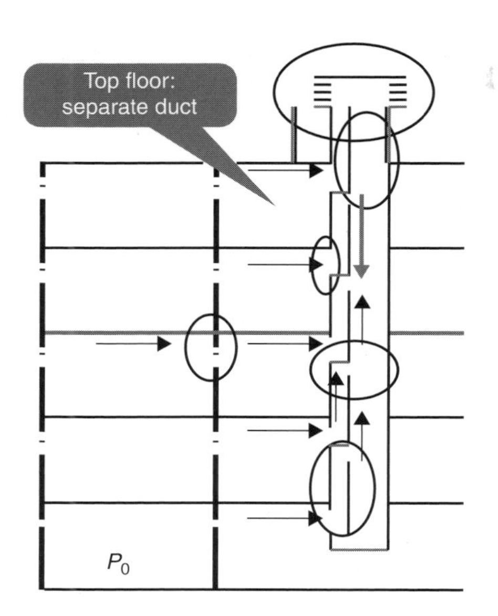

Figure 6.9 Typical "shunt duct" arrangement (*Source*: reference[17])

properly and that the tenants do not interfere with the extract terminals. With such a central system, the system will have to operate continuously. Therefore, there is the risk of excessive ventilation at times when moisture is not being produced. Probably the best course of action is to make use of humidity-modulated extract terminals. It is preferable that this is done in conjunction with matching air inlets. Such systems have been shown to give significant energy savings over unmodulated systems.

Given the complexity of the problems associated with the use of passive ventilation in buildings, there is a tendency to favour the use of centralised mechanical extract systems. The logical progression from this is to use some sort of heat-recovery system in order to reduce energy consumption. The large size of the overall ventilation requirement for a block of flats means that many of the scaling problems associated with heat-recovery units in individual dwellings are avoided (see Chapter 5). In order to make proper use of air-to-air heat recovery via centralised plant, supply air will be needed in flats as well as extraction. This further complicates the arrangements of ductwork. Room will also be needed for the central plant, usually at high level.

A duct system is potentially a means by which smoke and heat can pass through a building at great speed, and quite possibly from flat to flat. Beyond a maximum floor height above ground of 4.5 m, corresponding in most cases to more than two-storey height in the case of dwellings, regulatory restrictions start to apply. Suffice it to say at this juncture that fire dampers will be needed at any point where a duct leading from one flat penetrates the envelope of another flat. In addition, any duct used will have to satisfy certain requirements in terms of resistance to fire as laid down within the regulatory requirements. If the ventilation system is of the mechanical supply and extract type, then it should be capable of being shut down in the case of fire. A mechanical extract system might provide some minor assistance in the removal of smoke from the building. However, Approved Document B asks for an extraction rate of 10 air changes per hour. An extract ventilation system does not normally provide this amount of extraction, although extra fan capacity could be provided in case a fire takes place. It should be noted that if mechanical smoke extraction is installed, then the basement of the building must be fitted with a sprinkler system. Furthermore, the ductwork system should be capable of withstanding extract gas temperatures of 400°C for a period of at least 1 hour.

6.9 Noise control and ventilation

There are a number of possible noise sources which might have an adverse impact on the quality of life of the occupants of dwellings in the immediate proximity. The precise nature of the noise sources vary. Road traffic noise is the most common problem. It can be regarded as a line source of noise. The principal factors affecting the size of the noise problem are the hourly traffic flow along the road, the mean speed of the traffic and the proportion of the traffic in the form of heavy vehicles. The gradient of the road, the type of road surface, the terrain between the road and the point of reception, the angle of view of the road relative to the point of reception and the reflection of noise from nearby building facades also have an influence. Good worked examples of a number of appropriate calculation procedures are contained in the joint Construction Industry Research and Information Association (CIRIA)/BRE publication *Sound Control for Homes*.[19] If the nature of the site and/or the road traffic is thought to be so unusual that the calculation technique is not thought to be suitable, then recourse will have to be made to noise-level measurements on the site itself. When making calculations of noise levels due to road traffic, it is prudent to make an allowance for future growth in traffic loads, notwithstanding current political developments.

As the number of flights increases and airports expand, the issue of the noise from aircraft becomes more important. The noise from aircraft is intermittent, and of course emanates from above the ground. This means that its effects on dwellings are different to road traffic noise. One important consideration is that whereas the impingement of road traffic noise can be influenced to some extent by the design of a new development (e.g. by the use of barriers, such as earth banks and by the positioning of the new dwellings themselves) in the case of aircraft noise none of the factors influencing the noise level on the site can be affected by design. The magnitude of noise levels on a given site is determined by reference to maps of noise-level contours which are available from the airport authority in question. It is very difficult to best guess the impact of increasing numbers of flights.

The significance of noise from railways has probably diminished over recent years in line with the decline in fortunes of the railways themselves. This situation may not continue. Expansion of railway services is seen as an important part of current plans to modernise and integrate the national transport network. The opening of the Channel Tunnel has resulted in a steady and significant increase in railway freight traffic. In addition to any increases in railway traffic, until recently, rail track was involved in the disposal of large amounts of land which had been identified as being surplus to requirements. Some of this land, in close proximity to railway routes, has inevitably been used for housing purposes. The prediction of noise from railways is not a simple matter, although it can essentially be regarded as a line source in the same way as a road. A reasonable guideline is that if a site is within 30 m of a busy railway line, there is no screening of the site from the line. Vibration is more commonly found with railways than with roads. If a site is within 60 m of a railway line, then there is a risk of a vibration problem. A further complication is ground-borne vibration, which may in some cases result in structural damage. Ground-borne vibration is affected by the weight and speed of trains, and also by local geological conditions. In practice, problems with ground-borne vibration are difficult to resolve. This is particularly true if the problems have started after the construction of the dwellings as a consequence of a change in the traffic pattern on the railway line.

Industrial noise sources are extremely diverse in type and nature, and therefore prediction of noise levels on affected sites is difficult. It is usually the case that a site noise survey in accordance with the guidelines listed in BS4142 will be required.[20] Whereas it will be fairly straightforward to quantify the implications of industrial noise for an affected site prior to building the development, changes to operations on the industrial site might completely change the situation.

A rational solution to the danger of noise ingress might reasonably be considered to be to not build houses in areas of high risk. The responsibility for allowing development in any area, including those which may be associated with a risk of noise impingement from a nearby source, lies with the local authority, who must take into account issues of noise along with a wide range of other factors. The local authority would also have some responsibility for considering the noise implications of proposed changes in operations on industrial sites in cases where planning permission was required. Planning Policy Guidance Note 24[21] exists in order to give guidance to local authorities in England on how to use their planning powers in order to minimise the adverse effects of noise on new developments. The Document supercedes Department of Energy (DOE) circular 10/73, which for many years fulfilled the same purpose. The intention was for Guidance Note 24 to offer rather more information than DOE 10/73.

Guidance Note 24 gives guidelines for making planning decisions in noisy areas. Essentially there are four categories of noise exposure, classified from A to D; the threshold levels of which vary with the type of noise source. The recommendations for each category are summarised as follows:

- *Class A*: Noise need not be considered as a determining factor in granting planning permission, although the noise level at the high end of the category should not be regarded as a desirable level.
- *Class B*: Noise should be taken into account when determining planning applications and, where appropriate conditions imposed to ensure an adequate level of protection against noise.
- *Class C*: Planning permission should not normally be granted. Where it is considered that permission should be given, for example because there are no alternative quieter sites available, conditions should be imposed to ensure a commensurate level of protection against noise.
- *Class D*: Planning permission should normally be refused.

In practical terms this means that as far as design is concerned, the relevant classes are A, B and C.

The requirements for noise protection are different between night and day, as clearly if noise can enter a property at night then the sleep of occupants may be disturbed. Actual data is sparse. The World Health Organisation (WHO) Guideline[22] recommends that a night time noise level of less than 35 dB(A) is needed in order that the restorative process of sleep be preserved. The same document recommends that the general threshold for the avoidance of annoyance is 55 dB(A). These two figures are used to form the boundaries between NEC A and B. At higher boundaries, the figure quoted not only depends on whether night or day is being considered, but also what type of noise source is being considered.

Having established at what thresholds action needs to be taken in order to control noise ingress, it would be timely to consider the measures that could be taken. Of course, it should be appreciated that the issue of ventilation is one of the several factors to be taken into account when designing for protection against noise ingress. The mass of the building structure and the glazing used are all very important. The use of double and secondary glazing is well understood. The need for special ventilation strategies is on the other hand less well established.

The nature of the measures to be taken depends on the type of noise source and whether the affected properties are new built. The Land Compensation Act 1973 led to the issuing of the Noise Insulation Regulations 1975,[23] followed some year later by the Noise Insulation Amendment Regulations 1988.[24] The aim of this package of legislation is to compensate residents when road traffic problems arise which affect noise levels in the vicinity of existing buildings; for example, in cases where new roads are built in close proximity to an established housing estate. The measure used to determine whether the compensation mechanism is triggered is the A-weighted sound pressure level which is exceeded for 10% of the 18-h period between 0600 and 2400 h in the course of a normal working day. This is more conveniently abbreviated to $L_{A10,18\,h}$. If this figure is shown to exceed 68 dB, and there has been an increase of at least 1 dB from the previous situation, then residents in affected properties receive compensation in the form of a grant for the installation of double-glazing, supplementary ventilation systems and, in certain cases, for the installation of Venetian blinds and insulated or double doors.

For the purposes of this book, it is the supplementary ventilation provision which is of interest. A standard ventilation opening of any description is a potential source of ingress for noise from the outside. It is a relatively simple matter to produce a version of a standard wall-mounted air inlet which has appropriate sound attenuation properties, and with a little more difficulty to achieve the same performance for a trickle ventilator. In the case of extract devices for moisture-producing room, such as mechanical fans, these are intermittent devices, and are in any case rather noisy in operation themselves. For such devices, insulation of the connecting duct through the wall can easily be provided, although little could be done for a fan mounted in a window pane.

Unfortunately, all these efforts to control noise ingress are undermined by the other component of domestic ventilation, namely the rapid ventilation provided by the opening of windows. Window opening habits vary considerably between individual occupants. For most of the time some occupants will choose not to open the windows at all, but in other cases windows will be left open for the greater part of the day. In the summer, the opening of windows will be necessary in order to control internal temperatures. The implications are that in rooms facing the noise source (referred to in the Sound Insulation Regulations as a treatable room) on a site with a noise problem, window opening can only happen at the expense of a significant increase in noise ingress. This is perceived as an undesirable state of affairs.

The solution is to provide a means of rapid ventilation which will not serve as a potential source of noise ingress. The method specified in the Noise Insulation Regulations is by provision of mechanical ventilation units. These units are of a similar size to a small storage heater. The unit contains an acoustically treated fan-powered inlet, together with a treated passive air outlet. The speed of the fan must be adjustable between at least two settings: a lower value of 60 m^3/h and a maximum value of 255 m^3/h. The wording of the Noise Insulation Regulations means that in theory it is possible to separate the supply unit from the extract, although in practice this is very rarely done. Restrictions are placed on the self-noise produced by mechanical systems. When the powered inlet is on the minimum setting, the self-noise must be less than 35 dB(A), whilst at the higher setting it must be less than 40 dB(A).

In order to satisfy the requirements of the Noise Insulation Regulations a plastered solid brickwork wall into which a ventilation unit is fitted would be expected to achieve a performance specification as follows:

- R_m (average sound reduction index over the 16 dB octave bands between 100 and 3150 Hz): 50 dB;
- R_w (weighted sound reduction index): 54 dB;
- $R_{A(Traffic)}$ (outside–inside level difference): 47 dB.

Of these three parameters, the first two are measured in the laboratory, and are those most often quoted by manufacturers. Conditions within a laboratory may not necessarily be comparable to those within a real dwelling.

Planning Note 24 advances the perception of what constitutes an acceptable strategy for noise-free ventilation to a position slightly beyond that of the Sound Insulation Regulations 1975: even though it must be borne in mind that its recommendations only apply to new development. The need to keep closed windows and ventilators not designed to provide sound insulation during the period of risk is recognised. The important requirement for ventilation for summer time temperature control is already mentioned. Furthermore, it is also pointed out that some ventilators may provide a potential route for noise ingress even when closed, as a result of their design. Planning Note 24 cautions against the use of laboratory data for estimating the performance of a building element when installed in a property. Whereas reference is made to the use of the ventilator units described in the Noise Insulation Regulations, the Planning Note also mentions the use of "whole-house systems". The term "whole-house system" is not defined, but could reasonably be taken to mean a suitably treated PSV installation or installations complete with attenuating inlets, or else a ducted mechanical system with attenuators. It is stated that the use of ventilators in accordance with the Noise Insulation Regulations will restrict insulation against traffic noise to about 38 dB(A). This value is very slightly less than the $R_{A(Traffic)}$ quoted for compliance with the Noise Insulation Regulations, but the difference is insignificant.

The issue of what constitutes a satisfactory whole-house system is a vexed one in the absence of a defined performance standard for all possible systems. Recognising the limited validity of laboratory measurements, Edwards *et al.*[25] performed a programme of *in situ* tests of the sound attenuation capabilities of a range of PSV duct types and configurations on a housing estate falling under the requirements of the Noise Insulation Regulations as a consequence of its proximity to Birmingham International Airport. The measurements were based on the method described in the then prevailing BS test procedure, and involved measuring the transmission of noise at 16 dB octaves between 100 and 3150 Hz. Several interesting facts emerged. The first was that a PSV system with no special acoustic treatment had an intrinsic sound attenuation of between 20 and 25 dB at 500 Hz depending on the length of the duct being used. It was found that the use of attenuating ductwork gave a superior performance to the use of a silencer, the conclusion being that it would be appropriate to use a silencer only in cases where a long run of ductwork was involved. Extended monitoring of sound levels within a selected test property revealed no significant difference between internal noise levels measured when using acoustically treated ductwork compared to periods when the mechanical ventilator units were being used. In an area affected by aircraft noise, a PSV system discharging at ridge height is far more exposed to the actual source of the noise than would be the case with the wall-mounted mechanical units.

These findings indicate that the provision of acoustically treated PSV systems, complete with appropriate acoustically attenuated air inlets as to match the ventilation performance requirements of Approved Document F, gives a satisfactory acoustic performance. However, there is an important feature of the performance of PSV systems which serves to put a question mark against their acceptability as a direct replacement for the mechanical units. It must not be forgotten that the reason for the provision of the mechanical units was to compensate for the fact that windows could not be opened due to noise ingress problems; in other words, as an alternative means of rapid ventilation. In contrast, PSV systems constitute a method for the provision an enhanced level of background ventilation. Its extract performance declines during the warmer months as inside–outside temperature differences decrease from their winter levels. The particular reservation is that it will not be possible to deliver adequate air flow for purposes of controlling temperatures during the summer period. Research is in progress in which the performance of in-line booster fans within passive stack ducts is being monitored. These may be either switched on manually, or else may be controlled by means of a temperature sensor. This would mean that the passive stack duct would operate as a ducted extract ventilation system during the periods of high internal temperatures. This is an interesting development. An immediate concern is that the use of the fan will boost energy consumption compared to when the duct is operating only under the influence of natural-driving forces. Another issue is that of air velocities. It is almost certain to be the case that the air velocities induced within the occupied space by a boosted whole-house system will be much less than for window opening or for that matter than those induced by a mechanical unit. The argument could run one of the two ways. On one hand, reduced air velocities may result in ventilation not achieving as high a level of thermal comfort due to the absence of the perceived cooling sensation of draughts resulting from window opening. Equally, it might be argued that the elimination of gusting draughts might be favourably received. The results of this particular programme of research are awaited with some interest.

Before moving on from this topic, it would be useful to pause to consider a neglected yet most important aspect of the ventilation of dwellings in noise-affected areas, namely that of the behaviour of occupants. This is yet another aspect of building performance in which the interaction of occupancy with environmental systems has not been considered. Putting the issue basically, what assurance do we have that occupants will actually use mechanical units if they are installed, and if not, *what* do they do? As part of the study, in which the *in situ* measurements of noise transmission in passive stack ducts were made, Edwards *et al.*[25] also carried out a small-scale (about 60 questionaires) social survey of residents on the same estate, in which an attempt was made to quantify their habits in terms of the ventilation of their houses. The return rate on questionnaires was about 40%, which is fairly high for a survey activity of this type. The findings were very revealing. Over 50% of all respondents said that they never used the mechanical units at all, whilst over 60% said that they made use of window opening at some time of the year. During the summer months, the figure for use of window opening rose to over 85%. Only 10% of the respondents said that they never opened their windows. Notwithstanding the relatively small size of the responding group and the fact that the site was primarily affected by aircraft noise rather than the more prevalent problem of road traffic noise, there seems to be a clear indication that given the chance, many occupants will continue to open windows regardless of the external noise, with a larger proportion of them choosing to do so in the summer. These findings put question marks against both the need for mechanical units at all and also against the notion of making provision for extra ventilation air by mechanical means in anticipation of difficulties with the control of indoor temperatures during the summer season as a consequence of windows being opened. It is the opinion of the author that any deliberations over the provision of rapid ventilation in dwellings should include the most careful review of whether the perceived need does in fact exist.

A few words are necessary about noise control measures in buildings affected by noise from sources other than road traffic. In the case of aircraft noise, the qualifying level over a 16 hour period is essentially the same as for road traffic, as is the package of noise insulation measures offered. Planning authorities are responsible for setting out their own policies for dealing with planning cases where the site in question is affected by railway noise. As the noise nuisance from trains is much more intermittent than from road traffic or aircraft, the calculation procedure for determining the overall noise level is more complex, as it has to be time weighted. Those responsible for new railways will be obliged to provide sound insulation packages for the occupants of existing properties if an equivalent level of 68 dB is reached during the 18 hour period between 0600 h and midnight, or else if an equivalent level of 63 dB is reached between midnight and 0600 h.

Transmission of noise between spaces via ventilation ductwork is a problem which is more usually associated with systems within mechanically ventilated or air-conditioned building. However, difficulties may arise in dwellings. A good example would be in a block of flats served by a central mechanical system. For example, the ductwork serving a ground-floor flat may result in noise being transmitted to the flat above. In the case of a mechanical system, the recommendations given in Sections A1[26] and B5[27] of the CIBSE Guides should be followed in order to ensure that noise transmission via ductwork does not constitute a problem.

Having used the example of a mechanical system, it must be understood that passive stack ducts may also lead to noise transmission between rooms if precautions are not taken. At any point where a passive stack duct penetrates a wall or floor, a weak point is created in terms of sound penetration. In many cases, the element of the building envelope in question would have been designed so as to limit sound transmission. In such circumstances, it is important that appropriate measures are taken to ensure that noise breakthrough of adjacent properties does not take place. These measures should include the use of attenuated ductwork, or, where appropriate, silencers.

References

1 HW Harrison. *Roofs and Roofing – Performance, Diagnosis, Maintenance, Repair and the Avoidance of Defects* p. 54, Building Research Establishment, 1996 ISBN 1 86081 068.
2 The 1991 English House Condition Survey, HMSO, 1993.
3 RE Edwards, C Irwin. Ventilation of cold pitched roof spaces. Presented at the International Symposium on Roofs and Roofing in Bournemouth 1988.
4 BS5250 *Control of Condensation in Dwellings*. British Standards Institution, 1989.
5 RE Edwards, C Irwin. Multiple cell air movement measurements. *6th AIC Conference, Ventilation Strategies and Measurement Techniques*, The Netherlands, September 1985; pp. 8.1–8.18.
6 WH Ransom. *Building Failures – Diagnosis and Remedy*. E&F Spon, 1981. ISBN 0 419 11760 1.
7 UMIST Final Report for the Building Research Establishment, 1984 (not publicly available).
8 C Irwin, Private communication, 1987.
9 S Sharples. Eaves and roof ridge pressure coefficients on an isolated low rise dwelling: wind tunnel study. *Building Services Engineering Research Technology* 1997, 18(1), 59–61.
10 Department of the Environment. *Waste Management Paper 27: Landfill Gas*. HMSO, 1996.
11 *Approved Document C of the Building Regulations for England and Wales*, HMSO, 2000.
12 *Protective Measures for Housing on Gas-contaminated Land*. BRE Report 414, 2001, ISBN 1 86081 460 3.
13 *Thermal Insulation – Avoiding Risks*. BR 262, BRE/CRC, 2002.
14 *Radon: Guidance on Protective Measures for New Dwellings*. BRE Report 212, 1991, ISBN 0 85125 511 6.

15 *Non-traditional Housing in the UK – A Brief Review*. BRE and the Council of Mortgage Lenders, 2002.
16 *Limiting Thermal Bridging and Air Leakage: Robust Construction Details for Dwellings and Similar Buildings*. The Stationary Office, 2001, ISBN 0 11753 612 1.
17 AM Bernard, PJ Vialle, MC Lemaire, O Noel, P Barles. Monozone modelisation of natural ventilation with ducts. *Proceedings of the 22nd Annual AIVC Conference*, Bath, AIVC 2001.
18 P Wouters, D L'Heureux, B Geerinckx, L Vandaele. Performance evaluation of humidity controlled natural ventilation in apartments. *Proceedings of the 12th AIVC Conference, Air Movement and Ventilation Control within Buildings*, Ottawa, Canada, September 1991, Vol. 1, pp. 191–192.
19 *Sound Control for Homes*. BRE/CIRIA Report No. BR238 & CIRIA127, 1992.
20 *Method for Rating Industrial Noise Affecting Mixed Residential and Industrial Areas*. BS 4142:1997.
21 *Planning Policy Guidance 24 (PPG24) Planning and Noise*. Department of the Environment, HMSO September 1994.
22 *World Health Organisation Guidelines for Community Noise*, WHO, Geneva, March 2000.
23 *Noise Insulation Regulations*. HMSO 1975.
24 *Noise Insulation Amendment Regulations*. HMSO 1988.
25 RE Edwards, N Poole, A Jellyman. Acoustic Attenuation of Passive Stack Ducts. Final Report to the Department of the Environment, 1995.
26 Section A1, *CIBSE Guide A: Environmental Design*, CIBSE, 1999.
27 *CIBSE Guide B5: Noise and Vibration Control for HVAC*, CIBSE, 2002.

7

Regulations and Standards

7.1 Building Regulations

7.1.1 *Origins of regulation*

Before moving on to the actual regulatory requirements, it is useful to give a brief overview about the origins of the regulation of the construction of dwellings within the UK.

In theory, local authorities have had the power to exercise control of buildings since 1189. In practice, legislation covering the standards of construction were not necessary, given the relatively small population, and the lack of development of sanitary provision and other technical advances.

The Industrial Revolution was to change many things apart from the way that raw materials and goods were produced. One result of the increase in the number of factories was the emergence of very high concentrations of workers in the industrial cities. Overcrowding was found in commonplace, and much of the accommodation was of poor quality and with non-existent sanitary provision. Frederich Engels gives a particularly harrowing description of the appalling conditions in the slums of the Shudehill area of Manchester.[1] Modern-day visitors to the newly regenerated commercial developments and public open spaces of this part of Manchester see no evidence of this particularly dark era in the history of Manchester.

Looking back with the gift of 20th-century hindsight, it was inevitable that such conditions would lead to the spread of infectious diseases. Unfortunately, during the early 19th century, the links between poor housing, poor sanitation and disease were not fully established; indeed, it could be said that they were ignored. At this time, the prevailing view among the privileged classes was that disease was rife amongst the poor as a consequence of their moral failings. The priority of most of the wealthy people was to give money to the church in order to assure their safe passage to Heaven, rather than to give any relief to the lower classes. In fact, it is a small wonder that Marx and Engels thought that the slums of industrial England offered a fertile incubator for the Communist Revolution.

It took the appearance of cholera in Britain in 1831 to initiate any semblance of progress. Cholera spread at an alarming rate through society as a whole; its progress was not limited by the class divisions. This was a major blow to prevailing notions of social order. It became clear that measures had to be taken to try to limit its spread. The Cholera Act should be remembered not for its relative ineffectiveness in controlling the spread of cholera, but rather as the first occasion on which legislative powers were taken which permitted intervention within housing for the purposes of improving health. In the case of the Cholera Act, this intervention came in the form of powers of compulsory entry for purposes of the cleansing and fumigation of properties.

The recognition of the condition of housing as a determinant of health and ultimately longevity only took place in the middle of the 19th century as a result of the work of public health pioneers, such as Edwin Chadwick. Based on the developments from this recognition came the first attempts to produce pieces of legislation that were intended to improve housing standards. These pieces of legislation were pushed through Parliament, despite attempts to prevent such a course of events. The Public Health Act of 1848 did not seek to intervene directly in issues relating to housing. However, the Act did establish a centralised Board of Health, and was in effect the forerunner of what we would now recognise as the Department of Health. This Act also represented the first step towards the establishment of a *national code for public health and of standards for water supply and sanitation*.

The first Act of Parliament to have a direct bearing on housing was the Artisans and Labourers Dwellings Act of 1867. Under this Act, the Medical Officer of Health was empowered to order the demolition of housing that was deemed to be of unsatisfactory quality. The scope of the Act only applied to individual dwellings, and this was clearly a severe limitation on the usefulness of the Act, as it was extremely difficult to condemn and demolish large areas of slum housing *en bloc*. The Artisans and Labourers Dwellings Improvement Act of 1875 remedied this deficiency by sanctioning wholesale clearance of condemned properties.

Despite the aforementioned acts, the overall quality of housing among the poor continued to decline, and worse still. The demolition of unsatisfactory housing was not matched by the construction of replacement housing. This, combined with the rapid growth in population in the industrial cities, merely resulted in worse overcrowding within the rest of the housing stock. This was the most unfortunate outcome. It was soon realised that the government would have to widen its involvement in housing to encompass what we might perhaps now call the two ends of the dwelling life cycle. The first Act of Parliament, which actually introduced controls over the standards of new buildings, was the Public Health Act

1875. This gave local authorities statutory powers to enact bye-laws relating to such issues as ventilation, water supply and drainage.

After the enactment of the Public Health Act 1875, it soon became clear that despite the new powers available, progress was not being made in improving standards. This was quite simply because the costs involved in implementing improvements were inconsistent not only with the rents that the huge number of low-income tenants were able to afford, but also with the level of investment acceptable to landlords when compared to the return on that investment. To a landlord, the main attraction of slum housing was that they need not have to spend any money in order to make a good profit. Privately owned housing was essentially the preserve of the rich, whilst social housing was almost unheard of, despite the efforts of the so-called philanthropic housing movement to provide affordable rented accommodation of acceptable quality. (Even as late as the end of World War I, over nine-tenths of the British housing stock was under the control of private landlords.)

Eventually, it was accepted that the unfettered application of free-market principles was an impediment to the improvements of standards in housing, and that legislation had to be enacted in order to overcome the problem. The Housing of the Working Class Act 1890 gave local authorities the power to either build housing for let or else to obtain it by other means. These powers were extended by the House and Town Planning Act 1919.

The Housing and Town Planning Act 1919 was very important for two reasons. Firstly, the role of local authorities was changed to a statutory duty to provide sufficient amounts of suitable housing in the area under their jurisdiction. This in itself was a radical change. However, there was another change that was probably more important. Central government undertook the responsibility to provide funding for the provision of such housing, rather than allowing it to remain with the local authorities. This was seen as the most effective way to circumvent the profit motive.

Another significant event happened after World War I. The Tudor Walters Committee published its report on the design of housing and environment. The report was concerned with such issues as density of building, layout of estates and the provision of basic amenities, such as baths. Whilst not actually being enshrined in legislation, the standards set out in the Tudor Walters Report were adopted in the construction of the vast majority of the new council housing estates built as a consequence of the Housing and Town Planning Act 1919. As an added bonus, the standards also proved to be a major influence on the standards used for the construction of private housing.

As the 20th century has progressed, the overall purpose of the housing-related legislation has changed subtly. The infectious sanitation-related epidemics of the 19th century, which led to the establishment of the link between human health and quality of housing, are now extremely rare in the UK. If they do occur, they are more often than not brought home by tourists. As a result, after World War II, health alone was no longer the primary driving force for improvements in housing and developments in legislation.

In 1961, the report entitled *Homes for Today and Tomorrow* was published. This report was produced by a Committee chaired by Sir Parker Morris.[2] It considered standards for new housing, and included a set of design standards for housing which went above and beyond the level of any previous notion of what was considered to be an acceptable standard. Despite the undoubted merits of the so-called Parker Morris Standards, they were never incorporated in their entirety into any mandatory design code. However, rules were introduced whereby central government would only provide the requisite amount of financial subsidy for the construction of council houses if compliance with the Parker Morris Standards was achieved. As a result, most of the new council houses complied with the Standards, and a big step forward was taken. Later in time, compliance with the Standards was made all but mandatory for the New Town Corporations in 1967, whilst compliance was made compulsory for all local authorities in 1969. With the introduction of national building regulations, this compulsion was removed. However, the Parker Morris Standards survived until the early years of the Thatcher Government, when they were formally abolished as part of the relentless drive towards deregulation. Looking at the dwellings being currently built, it would seem that the single biggest change arising from the demise of Parker Morris Standards has been the progressive reduction in the size of dwellings and the rooms therein.

Up until 1965, there was no national building regulations framework. All local authorities had their own regulations. After 1965, the Building Act meant that there was a national framework of uniform regulations. These developed slowly up until the introduction of the Building Act 1984. Changing the regulations was a time-consuming business.

The current situation regarding Building Regulations is not straightforward. The prevailing Regulations are dependent on the geographical location. Three separate documents exist for England and Wales, Scotland and Northern Ireland. The arrangement of the Regulations is essentially the same in all cases.

The Building Act 1984 sets out the current framework for Building Regulations. According to the 1984 Act, Building Regulations may be made for three basic reasons:

1. For securing the health, safety, welfare and convenience of people in or about buildings and of others who may be affected by buildings or matters connected with buildings.
2. For furthering the conservation of fuel and power.
3. For preventing waste, undue consumption, misuse or contamination of water.

This represents a very wide remit, and indeed this is reflected in the scope and coverage of the Approved Documents. However, it is interesting to note that in their current form, the Building Regulations are now probably as far away as they have ever been from representing a definitive set of standards for housing.

The means of enforcement for Building Regulations is via building control. In 1985, the system of building control in England and Wales was subject to major changes. These came into effect in November 1985. With some exemptions, all building work in England and Wales falls under the Building Regulations.

The prevailing set of Regulations for England and Wales are in the format prescribed in the Building Regulations 1991. These came into force on 1 June 1992. Amendments were made in 1994. The basic Building Regulations themselves are simple statements of requirement, with no technical information related to compliance. The technical details of recommendations that should be followed in order to achieve compliance are contained within a set of 14 Approved Documents. (The idea of Approved Documents was first introduced in the 1984 Act.)

By chance, Approved Document F, which is of most interest to readers of this book in the context of domestic ventilation, was one of the first to be revised under this process. The revised Approved Document, together with a revised Approved Document L (Conservation of Fuel and Power) and Approved Document A (Structure) came into law as the Building Regulations (Amendments) 1994. An important point to bear in mind when considering any Approved Document is that Approved Documents are not actually mandatory documents, merely a collection of possible means of achieving Building Regulations compliance. In many cases, there will be other means of compliance, often involving reference to British Standards (BS) and similar codes. These will be listed within the Approved Document but their recommendations will not be described. In the case of ventilation, other means of compliance will be described later in this chapter. There is a good selection of very helpful books concerned with the interpretation of the Approved Documents. The one most frequently consulted by the author is that written by Powell Smith and Billington, and more recently by Billington *et al.*[3] Bob Waters will be remembered amongst experienced ventilation researchers for his involvement in tracer-gas measurements in large spaces in the 1980s and 1990s.

The introduction of the Approved Documents arrangement was intended to give greater flexibility for revisions to particular areas of coverage in the Building Regulations, so that only the relevant Approved Document needs to be revised, rather than the entire set of Regulations. Flexibility of amendment from the viewpoint of economy of Parliamentary time has, undoubtedly, been achieved. However, it is highly questionable whether the same could be said of the actual process by which any revisions are proposed, reviewed, amended and finalised. It seems that a far greater number of organisations are now involved in the consultation process. These include academic institutions, professional bodies (such as the engineering institutions), trade associations, charities and lobbying groups. There have been two notable examples of protracted lobbying against particular proposals, namely the specific mention of passive stack ventilation (PSV) in the 1995 version of Approved Document F and against measures intended to reduce the use of air conditioning in the 1994 version of Approved Document L.

In the case of the Building Regulations for England and Wales, there are clauses relevant to the aspects of domestic ventilation contained within four distinct sections:

- *Part C* contains recommendations for the control of radon and methane.
- *Part J* covers issues related to the supply of adequate amounts of combustion air to gas appliances.
- *Part N* (introduced in 2000) places restrictions on the design of opening windows which whilst probably intended for application to non-domestic buildings should be borne in mind when contemplating windows within the dwellings.
- The current version of *Part L*, whilst not explicitly making regulatory demands for the dwellings, encourages the application of good practice in detailing so as to reduce air infiltration through the building fabric in commercial buildings and backs this up with a performance standard.

7.2 Building Regulations for England and Wales: Approved Documents relating to ventilation

7.2.1 *Approved Document C*

In terms of scope of coverage, Approved Document C of the Building Regulations is probably one of the most diverse of the Approved Documents. Its full title is *Materials, Workmanship, Site Preparation and Moisture Exclusion*. Only small amount of the content of the document has any direct bearing on ventilation provision within the dwellings themselves. The exception to this is Section C2, which gives provisions relating to the sites containing dangerous and offensive substances (see Figure 7.1). The precise Regulation stipulates that:

> Precautions must be taken to prevent any substances found on or in the ground causing a danger to health and safety.

For purposes of Approved Document C, the term contaminant is defined as:

> ... any material (including faecal or animal matter) and any substance which is or could be become toxic, corrosive, explosive, flammable or radioactive and therefore likely to be a danger to health or safety. This material must be in or on the ground to be covered by the building.

Signs of possible contaminants	Possible contaminant	Relevant action
Vegetation (absence, poor or unnatural growth)	Metals Metal compounds*	None
	Organic compounds Gases	Removal[1]
Surface materials (unusual colours and contours may indicate wastes and residues)	Metals Metal compounds*	None
	Oily and tarry wastes	Removal, filling or sealing
	Asbestos (loose) Other mineral fibres	Filling[2] or sealing[3] None
	Organic compounds including phenols	Removal or filling
	Combustible material including coal and coke dust	Removal or filling
	Refuse and waste	Total removal or see guidance
Fumes and odours (may indicate organic chemicals at very low concentrations)	Flammable explosive and asphyxiating gases including methane and carbon dioxide	Removal
	Corrosive liquids	Removal, filling or sealing
	Faecal animal and vegetable matter (biologically active)	Removal or filling
Drums and containers (whether full or empty)	Various	Removal with all contaminated ground

Notes:

Liquid and gaseous contaminants are mobile and the ground covered by the building can be affected by such contaminants from elsewhere. Some guidance on landfill gas and radon is given in Approved Document C; other liquids and gases should be referred to a specialist.

* Special cement may be needed with sulphates.

Actions assume that ground will be covered with at least 100 mm *in situ* concrete.

[1] The contaminant and any contaminated ground removed to a depth of 1 m below lowest floor (or less if local authority agrees) to place named by local authority.

[2] Area of builiding covered to a depth of 1 m (or less if local authority agrees) wtih suitable material. Filling material and ground floor design considered together. Combustible materials adequately compacted to avoid combustion.

[3] Imperforate barrier between contaminant and building sealed at joints, edges and service entries. Polythene may not always be suitable if contaminants are tarry waste or organic solvent.

Figure 7.1 Risks from contaminated sites (taken from Billington *et al.*[3])

A list of sites likely to contain contaminants is given as Table 1 of the Approved Document C2: this is given as Table 7.1.

As pressure for building land increases, the need to use sites of the types given in Table 7.1 will grow. Two well-known developments on affected

Table 7.1 Sites likely to contain contaminants (taken from Approved Document C2, 1991)

Asbestos works
Chemical works
Gas works, coal carbonisation plants (coking) and ancillary by-product works
Industries making or using wood preservatives
Landfill and other waste disposal sites
Metal mines, smelters, foundries, steelworks and metal-finishing works
Munitions production and testing sites
Nuclear installations
Oil storage and distribution sites
Paper and printing works
Railway land, especially larger sidings and depots
Scrap yards
Sewage works, sewage farms and sludge disposal sites
Tanneries

sites are the Millennium Dome and the Manchester Commonwealth Games Stadium. These are both built on the sites of former gas works. However, the issues involved with these developments relate to the remediation of contaminated land, and this is in fact the case with the majority of the sites listed in Table 7.1. The technology of contaminated land remediation is beyond the scope of this book, and will not be dealt with here. From the list (Table 7.1), landfill sites pose the clearest risk to dwellings in the form of potential sources of landfill gases.

Landfill gas is encompassed by the definition of a contaminant given in Approved Document C2, as another risk not linked to previous site usage, but rather to the geology of the sub-site. The nature of these two risks and their influence on occupant health, together with possible strategies for control, are discussed in some detail in Chapters 2 and 7, respectively. Section 7.2.1 will concern itself with the regulatory requirements for demonstrating compliance Regulation C2.

Approved Document C gives specific advice regarding the siting of new buildings in relation to a landfill site that may be a source of harmful gases. It is stated that if any piece of ground to be covered by a building is on, or is within 250 m of a landfill, or if there is reason to suspect that there may be gaseous contamination of the ground, or that the proposed building or buildings will be within the likely sphere of influence of a landfill, then further site investigations should take place. However, the Approved Document makes no recommendations regarding practical means to control the ingress of landfill gases into the dwellings. Instead, the reader is recommend to refer to Building Research Establishment (BRE) Report BRE212 for design guidance.[4]

With respect to the control of indoor radon gas levels, the reader is requested to refer to BRE Report BRE211:[5] little guidance is contained within the Approved Document itself. On the basis of the tables provided in BRE211, a decision must be made as to whether full radon protection (i.e. primary measures in the form of a radon-proof barrier plus secondary measures in the form of a radon sump plus extract system or a ventilated sub-floor) or merely secondary measures are required to keep indoor radon concentrations below the 200-Bq/m^3 action level. The main reason for this relies on a BRE Report, which states that number of sites falling under the requirements of Approved Document C is increasing as the National Radiological Protection Board (NRPB) makes progress through its programme of measurements at sites with geological details indicative of possible problems with the emission of high levels of radon gas. Updating BRE211 is seen as a more convenient updating process than having to carry out repeated revisions to Part C. Even then, there will be a delay between the NRPB reporting on its findings and BRE211 being updated. Therefore, it is recommended that the local building control authority is contacted to confirm the current status of the lists of affected areas prior to the planning of any construction activity on a potentially affected site.

Part C4 deals with the prevention of moisture ingress via suspended floors, both of concrete and of timber, by both constructional detailing and by the provision of ventilation. The ventilation requirement is met by the provision of ventilation openings in two opposing external walls which allow ventilation to all parts of the floor. Although not mentioned in the Approved Document, complete ventilation of a compartmentalised sub-floor can be achieved by leaving gaps in the sleeper walls. Openings should be provided at a rate of 1500 mm^2/m run of wall. In some cases, air may have to be ducted into the sub-floor cavity. In such cases, the ducting used must have a diameter of at least 100 mm. The depth of a sub-floor cavity should be at least 150 mm. This is in reality a very small space, certainly too much small to permit under-floor access for the purposes of replacing pipes and cables.

It appears that major restructuring of Part C is being considered by Department of the Environment, Transport and the Regions (DETR). These are discussed in more detail in Chapter 8.

7.2.2 *Approved Document F*

It is within Part F and its accompanying Approved Document that the principal requirements for the ventilation of the dwellings are laid down. The actual regulation F1 itself is very simple. It states that:

> there shall be adequate ventilation provided for people in dwellings.

There is another regulation, F2, which states that:

> adequate provision must be made to prevent excessive condensation in roofs and roof voids over insulated ceilings.

Clearly these two clauses does not provide any information about the actual means that may be used in order to achieve compliance. This lies within the Approved Document itself. The most recent version of Approved

Document F came into force in July 1995. There does not appear to be any other part of the Building Regulations for England and Wales that offers so many alternative means of compliance. Prior to the introduction of the most recent amendments, originally, Approved Document F was only concerned with ventilation in the dwellings. However, the 1995 version has been extended to cover non-domestic buildings. Although not of direct relevance in the context of this book, it should be noted that in the case of non-domestic buildings, compliance with the requirements of Approved Document F would effectively prevent the serving of an improvement notice under Section 23(3) of the Health and Safety at Work Act 1974, with relation to the requirements for ventilation provision as given in Regulation 6(1) of the Workplace (Health Safety and Welfare) Regulations 1992. This is noteworthy in that it is the first time that such a linkage has been made between the Building Regulations and legislation relating to health and safety in the workplace. It is very likely that other such linkages will be developed, as other matters relating to non-domestic buildings find their way into the Building Regulations. Indeed, this is probably a rational strategy to pursue.

At the outset, it should be understood that there are several types of buildings or spaces that are exempted from the requirements of Regulation F1. These are:

- *garages* that are used only in connection with one specific dwelling (as opposed to communal garages such as might be found attached to blocks of flats);
- *spaces or buildings* that are used solely for storage purposes (some householders, the author included, might comment that their garage also falls into this category!);
- *buildings or spaces* where people do not normally go.

Within the domestic context, exemptions covered by the latter must be extremely rare. With respect to garages, Figure 7.2 serves to show that it is perfectly possible to produce mould growth in a garage not covered in Approved Document F. In this case, the problem was due to a tumble dryer discharging over a steel-roof beam. It must also be pointed out that contained within previous versions of Approved Document F there was a requirement to provide ventilation in communal areas within flats and controls over the use of enclosed courtyards. These were removed from the 1995 version, following the publication of some research (which incidentally included some very impressive field measure) by Woolliscroft and colleagues of the BRE[6] that demonstrated quite clearly that such provision was unnecessary. With hindsight, one wonders how the requirement managed to find its way into the earlier versions of the Regulations in the first place.

As far as the ventilation of dwellings is concerned, Approved Document F clearly settles on the provision of ventilation strategies for the control of

Figure 7.2 Mould growth on steel lintel in a garage

condensation and mould growth as its key priority. Special definitions of particular rooms needing appropriate ventilation provision are given within the Approved Document. These rooms are defined below.

- A *utility room* is defined as a room in which water vapour is likely to be produced in significant quantities because it is designed to be used to contain clothes washing or similar equipments. One important proviso is that a utility room is exempted from any requirement for ventilation provision if the sole means of access to the room is from the outside of the dwelling, which of course would mean that there would be no danger of any water vapour generated spreading through the occupied space if not removed. In the majority of modern new-built houses, the utility room will be accessible from the inside of the dwelling. Indeed, in the smaller houses being built today, the provision of a utility room is seen as a valuable feature to compensate for lack of space in other rooms.
- A *habitable room* is a room used for dwelling purposes which is not solely a kitchen. Thus, a kitchen/dining room area would count as a habitable room.
- An *occupiable room* is defined, for purposes of dwellings, as rooms occupied by people. The definition specifically excludes bathrooms, sanitary accommodation and utility rooms.

- A *bathroom* is defined as a room that contains a bath or shower. The bathroom could be either with or without sanitary accommodation.
- For domestic purposes, *sanitary accommodation* is defined as a room containing a water closet.

Approved Document F identifies the following levels of ventilation provision that could be used in various combinations within different rooms in order to give compliance of:

- *background ventilation* to ensure a minimum level of fresh-air supply in order that residual water vapour is dispersed within all the occupied spaces within the dwelling;
- *extracting moist air* at the source of production (typically within the kitchen or bathroom), which can be either mechanical or natural;
- *rapid ventilation*.

Background ventilation is usually provided in the form of trickle ventilators. The required level of provision is set at either 8000 mm^2 in moisture-producing areas, such as kitchens and bathrooms, plus 4000 mm^2 in other habitable rooms, or else at an average of 6000 mm^2 over all the habitable rooms in the dwelling. Any ventilator used must be adjustable, and it is recommended that they should be installed at a height of not greater than 1.8 m. It used to be the case that cheap trickle ventilation could be provided simply by routing out a slit in the wooden window frame, as shown in Figure 7.3. Demands for controllability and the prevention of the ingress of vermin mean that there has been a move towards aluminium and more recently plastic products as a superior material of construction. It has become very common for the ventilators to be delivered to the site as readily mounted in the window frames.

The issue of trickle ventilator size was a vexed question during the consultation period prior to the adoption of the 1995 Revisions to Approved Document F. In practice, 4000 mm^2 is the maximum open area which can be used in a trickle ventilator which will fit into a typical window frame of the type currently used. An opening of 8000 mm^2 in such a window frame would be so large as to be very difficult to fit, and would in any case compromise the strength of the window frame. The production of tooling in the manufacture of plastic for a new trickle ventilator size would be extremely expensive, and holding stocks of two different components was at time thought to be uneconomic. The 6000 mm^2 average clause permits compliance by the use of multiples of 4000 mm^2. In cases where the design of the specified window frames does not permit the use of trickle ventilators, there are two alternatives. The so-called *over-bar ventilator* could be specified. These are mounted above the glazed unit within the window area itself. Whilst providing a satisfactory means of compliance, over-bar ventilators may not be pleasing to the eye. Alternatively, a ventilation opening of the required open area could be provided in the wall

Figure 7.3 Routed slit in window frame for trickle ventilation

itself. There are several available ranges of products for this purpose. *Airbricks* are a permissible means of providing background ventilation, providing that any slot within an airbrick is at least 5 mm across, and that any circular or square holes are at least 8 mm across. These requirements exclude insect screening or baffles.

The provision of background ventilation within areas of high moisture production was only introduced into Approved Document L in 1995. The previous practice, as enshrined in the 1989 version of the Approved Document, was for no background ventilation to be provided in such rooms, reliance being placed instead upon rapid extraction by means of a fan. The intention of this was to promote the flow of air into such rooms when the extract ventilation was operating, thus reducing the risk of moist air migrating into other rooms by means of two-way air flows through the doors. It is not absolutely clear why the amendment was made. It is evident that an increase in ventilation should be expected; however, the danger of short circuiting of incoming fresh air directly to the extract device is not mentioned as a possibility if ventilation openings are not sited carefully, and in fact no guidance is given as to appropriate location of fans.

There are two principal means by which the *extraction of moist air* from areas of production can be achieved. Mechanical extract fans remain a popular choice, despite clear problems with maintenance and performance. Fans may be switched on and off manually, or else by some sort of controller; for example, one based on a humidity sensor. PSV is increasing in popularity, despite an incomplete understanding of the optimal design and installation parameters for such systems. With respect to PSV design, the reader is referred to BRE Information Paper IP13/94[7] for guidance on designs which will achieve compliance. Apart from the following design recommendations given within the Information Paper, it is stated that other system designs may be used provided it can be demonstrated that they will allow an adequate amount level of ventilation for condensation control. This would appear to leave two options for the demonstration of performance level, either computer simulation or site performance monitoring of trial systems. It is the author's experience that the latter course of action has thus far been the preferred one. Approved Document F makes mention of following the recommendations within BS5250,[8] discussed in more detail later, as an alternative means of compliance. This somewhat muddies up the water, as certain aspects of BS5250 1989 are out of date, and indeed a revised version of BS5250 was issued in 2002, so technically that the alternative means of compliance is taken from an out-of-date BS. No conflicts are apparent in the 2002 version of BS5250,[9] and so this particular avenue for demonstrating compliance seems to have been closed, even though the wording of the Approved Document lags behind.

A third means of providing extract ventilation is via a suitable open-flued heating appliance. This is because such appliances take air for combustion from the space in which they are installed, and may indeed induce a buoyancy-driven flow. However, it is important that such appliances should still provide adequate extract ventilation when they are not operating. The majority of solid-fuel-burning open-flued appliances are suitable for the purpose of providing extract ventilation, with the proviso that they are the primary source of space heating, cooking or hot-water production within the dwelling. Care should be taken when considering appliances which burn other fuels. These may have control dampers which prevent air flow through the flue when the appliance is not in use. Such appliances would not be suitable. In other cases, whilst no damper may be present, the flue diameter may be inadequate for extract ventilation purposes. The flue on an appliance should be at least the equivalent of a 125-mm diameter circular duct. If an open-flued appliance within a space requiring extract ventilation is not in fact suitable for that purpose, then it may well be the case that an extract device will have to be provided. If this device is mechanical, then the greatest care must be taken in order to ensure that the open-flued appliance runs safely when the extract fan is running, as there is a danger that the change in pressure regime may result in the spillage of combustion products into the space. Mechanical extract ventilation should never be provided in the same room as an open-flued solid-fuel appliance. However, a gas appliance should be safe, provided that the extract rate of the mechanical device does not exceed 20 l/s. No advice is given about open-flued oil-burning appliances; instead, the reader is directed to OFTEC Technical Note T1/112.[10] A combustion product spillage test, in accordance with BS5440 Part 1,[11] should be performed as a safety check. If spillage is shown to take place, then the extract rate should be reduced to a point at which it can be seen that spillage no longer occurs. It should be noted that a spillage test should also be carried out even if the appliance is in a different room, as the pressure differential induced by the extract may in certain cases still cause spillage. This is a timely juncture at which to remember that Approved Document J gives recommendations for the supply of combustion air to fuel-burning appliances.

The only realistic way of providing *rapid ventilation* is by the use of window opening. The stipulated figure is that the equivalent of 1/20 of the total floor area of the dwelling must be available as opening windows. There are other parts of the Building Regulations, which may impinge on the choice of windows for the provision of rapid ventilation. In particular, there may be a conflict between the provision of rapid ventilation and the provision of means of escape in the event of fire. The reader is recommended to study Approved Document B Section 1.18 for more information.

The provision of opening windows may well be a security hazard in certain areas, particularly at ground-floor level.

An alternative to the provision of ventilation from areas of moisture production is the use of a ducted mechanical ventilation system at a ventilation rate of 1 air change per hour, in accordance with the design recommendations given within the BRE Digest 398.[12]

7.2.3 *Requirements in individual rooms*

Approved Document F1, Sections 1.1, 1.2a and 1.2b state that kitchens, habitable rooms, utility rooms, bathrooms and sanitary accommodation should be provided with both background and rapid ventilations. These requirements have already been discussed in Section 7.2.2, and are also summarised for convenience in this section. In addition to this, kitchens, bathrooms and utility rooms, and bathrooms in domestic buildings should also be provided with extract ventilation, again in accordance with the requirements listed in Figure 7.4.

Room without an external wall cannot, of course, have an opening window to outside. Such rooms are increasingly common in new dwellings as every attempt is made to maximise space utilisation. Special requirements exist for non-habitable rooms which do not contain an opening window to outside. In such cases, Paragraph 1.5 of Approved Document F1 stipulates that extract ventilation should be provided by means of an extract fan of appropriate extraction capability, a passive stack duct, or else by a suitable open-flued heating appliance. In addition, an air inlet must be provided in order to allow a satisfactory feed of air to the extract

Ventilation recommendations for rooms capable of containing openable windows				Ventilation recommendations for rooms not containing openable windows	
1 Room or space	2 Room ventilation	3 Background ventilation	4 Extract ventilation (fan rates)	5 Mechanical extract ventilation (fan rates)	6 Notes
Domestic buildings					
Habitable rooms	Ventilation opening equal to at least 1/20th room floor area	8000 mm^2	See note column 6	For mechanical ventilation see BS5720 1979 and BRE Digest 398	No recommendation given in Approved Document F1
Kitchens	Opening window (any size)	4000 mm^2	30 l/s in or adjacent to hob. 60 l/s elsewhere	Mechanical extract as column 4, with 15 min overrun on fan connected to light switch for rooms without natural light	See also text page 11.7 and note at foot of column 4 below for alternatives to mechanical extract
Utility room	Opening window (any size)	4000 mm^2	30 l/s	Mechanical extract as column 4, with 15 min overrun on fan connected to light switch for rooms without natural light	No ventilation provisions necessary if room entered only from outside
Bathroom	Opening window (any size)	4000 mm^2	15 l/s	Mechanical extract as column 4, with 15 min overrun on fan connected to light switch for rooms without natural light	Bathroom may or may not contain WC
Sanitary accommodation if separate from bathroom	1/20th room floor area as habitable room in above	4000 mm^2	See note column 6		See also BS5720: 1979 and BRE Digest 398

Figure 7.4 Summary of Approved Document F ventilation requirements for rapid and background ventilation

system. A convenient way of achieving this is by leaving a 10-mm gap underneath the door, although other more elaborate methods, such as transfer grilles in doors and louvres above the doors, may also be used. The author is aware of several cases involving social-housing projects where louvres above doors have been discounted as an option following the advice about the fire implications. Whilst hearing the concerns, the logic is hard to discern, especially when internal doors for domestic use are constructed neither with fire resistance nor with airtightness in mind. Glass louvres have another advantage in that they permit the spread of daylight.

Special provisions for habitable rooms with no external walls are also described in the Approved Document. It is permissible for an internal habitable room to be ventilated through a connected adjoining room, provided that there is a permanent opening between them, which is of an open area equivalent to at least 1/20 of the combined floor areas of the two rooms. If this requirement is met, then the two rooms can be treated as one, and ventilation provision can be made in accordance with the recommendations given in Figure 7.4. However, if the requirement cannot be met, then mechanical ventilation should be provided. The Approved Document does not actually make a recommendation of how much air flow.

In some cases, a habitable room may open into a conservatory. This is a common occurrence when a lean to conservatory is added to the side of a house. It might be the case that the conservatory might interfere with the ventilation of the habitable room beyond it. In such circumstances, the Approved Document allows for the habitable room and conservatory to be treated as a single room, provided that the opening connecting the two spaces is of an area at least 1/20 of the combined floor areas. This is the same as the treatment of an internal room. However, in the case of the conservatory and internal room, it is stipulated that at least some (but not actually how much) of this open area must be 1.75 m above floor level. In addition to this, 8000 mm^2 of background ventilation should be provided between the internal room and the conservatory.

7.2.3.1 Alternatives to the Approved Document F1 recommendations

As has been noted earlier in this chapter, Approved Documents are not mandatory in nature. The government of the day was very keen on being seen to promote freedom of choice in many matters of regulation. In fact one might wonder why they continued to bother with Building Regulations in their existing form. In the case of Approved Document F, several alternative means of achieving compliance are specifically sanctioned.

7.2.3.2 The use of continuous mechanical ventilation

Requirements for the performance of continuous mechanical systems in dwellings are not described in Approved Document F. Instead, the reader is referred to the following list of clauses taken from BS5720[13] *Code of Practice for Mechanical Ventilation and Air Conditioning in Buildings*: Sections 2.3.2.1, 2.3.3.1, 2.5.2.9, 3.1.1.1, 3.1.1.3 and 3.2.6.

7.2.3.3 Other means of compliance

It is possible to achieve compliance by reference to other documents cited in the Approved Document F. The two most well known of these are the following:

1. BS5250 1989 *Code of Practice for the Control of Condensation in Buildings*,[8] (which has been superseded by BS5250 2002[9] and is in any case discussed in detail elsewhere in this chapter).
2. BS5925 1991 *Code of Practice for Ventilation Principles and Designing for Natural Ventilation*.[14]

In the case of BS5250 1989, Approved Document F deems the most relevant clauses to be 6, 7, 8, 9.1, 9.8, 9.9.1, 9.9.2 and 9.9.3, and Appendix C, whilst in the case of BS5925, the most relevant clauses are deemed to be 4.4, 4.5, 4.6.1, 4.6.2, 5.1, 6.1, 6.2, 7.2, 7.3, 12 and 13.

Alternatively, compliance may be achieved by following the recommendations given in BRE Digest 398 *Continuous Mechanical Extract Ventilation in Dwellings: Design Installation and Operation*. BRE Digest 398[12] covers two possible methods for continuous provision, one dealing with the whole house ventilation using balanced supply and extract, the other being confined to the ventilation of areas of high moisture generation and sanitary accommodation. In conventional dwellings, a balanced supply and extract system is a rare thing indeed, but ducted extract is encountered a little more often. A common practice is to provide centralised extraction for a network of ducts in cases where PSV would be inappropriate; for example, where a roof terminal cannot be positioned in a suitable place. If the fan is of appropriate duty, then its capacity to extract should not normally be adversely affected by the pressure regime around the discharge point.

7.2.4 *Regulation F2*

Regulation F2 deals with the ventilation of roof spaces.

It is worth reminding the reader that F2 states that:

> adequate provision must be made to prevent excessive condensation in roofs and roof voids over insulated ceilings.

A roof construction having a void over an insulated ceiling will often be referred to as being of cold construction, especially if the roof is of sufficiently low pitch as to be classified as a flat roof.

The Approved Document enlarges on the Regulation by recommending that under normal conditions, condensation in roofs and in spaces above insulated ceilings should be limited such that the thermal and structural performance of the roof will not be permanently and substantially reduced. The wording of the Approved Document, therefore, recognises the virtual impossibility of preventing roof-space condensation under conditions of extreme cold and acknowledges that the roof structure is in fact unaffected by transients in condensation conditions. Recommendations for the ventilation of pitched-roof spaces are described in some detail. The recommended provision of ventilation in duo-pitched roofs can be summarised as follows.

For roof pitches greater than 15°, the equivalent of a continuous permanent 10-mm opening around the external perimeter of the roof space should be provided in order to promote cross-flow ventilation, as shown in Figure 7.5. This should be situated at low level. Care must be taken not to obstruct the vents with roof insulation. For a mono-pitch roof of greater than 15° pitch, high-level ventilation equivalent to a 5-mm continuous opening should be provided at the highest possible position on the pitch, as shown in Figure 7.6. In some pitched-roof constructions, the insulation may follow the line of the roof pitch. An example of this would be in the case where a loft had been converted into an occupied space. Where such a construction detail has been used, special arrangements must be made to ensure adequate roof-space ventilation. The low-level opening must be equivalent to a continuous 25 mm gap instead of 10 mm. A 50 mm wide

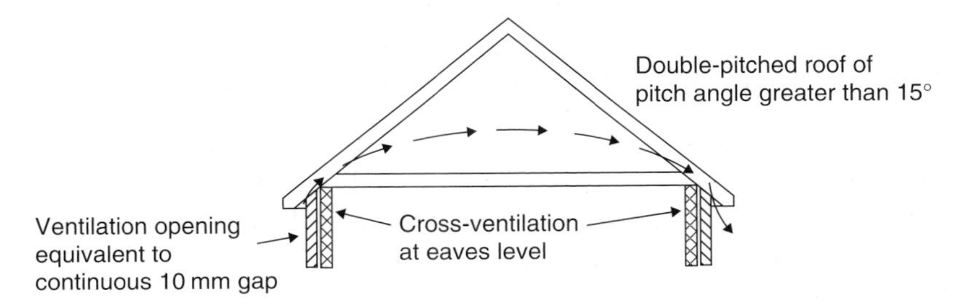

Figure 7.5 Roof ventilation greater than 15° (double-pitched roof)

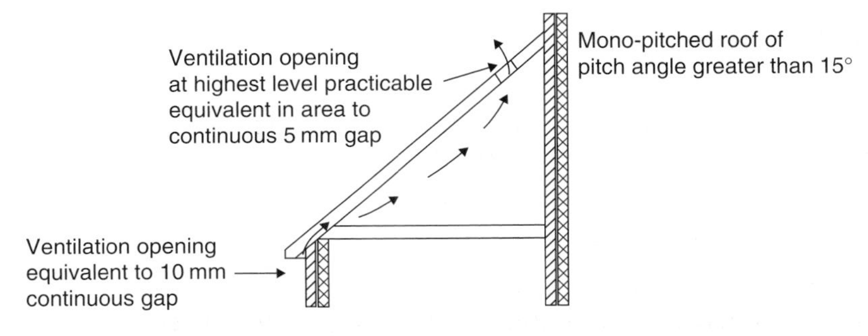

Figure 7.6 Roof ventilation greater than 15° (mono-pitched roof)

air space should be provided between the underside of the roof covering and the insulation, and high-level ventilation equivalent to a 5-mm continuous opening should also be used. This must be done in order to encourage the flow of ventilation air, otherwise a serious condensation problem will occur. Loft conversions are currently very much in vogue. One cannot help but wonder whether the need for unimpeded air flow through the cavity is fully understood by all those carrying out the conversion work.

For roof pitches less than 15°, it would at first seem that it would be much easier to achieve full cross-flow ventilation of roof space. In fact, the risk of condensation is greater, because the volume of the roof space is smaller and, therefore, becomes saturated with water vapour more rapidly. This clause of Approved Document F2 effectively sets out requirements for what would probably be referred to by many people as "flat roofs". Such roofs have historically been bedevilled by failure in service, and whereas inadequate ventilation has been a contributing factor, there are several others which are at least significant. In the case of such a roof space, the equivalent of a continuous 25-mm opening around the external perimeter of the roof space should be provided, again in the form of permanent vents, as shown in Figure 7.7. A free air space of 50 mm width should be provided between the insulation and the roof deck. If the roof joists run normally to the intended direction of air flow through the roof, then counter battens should be used to form the free air space. Special provisions are laid down for roofs where the span is in excess of 10 m, or when the plan layout of the roof space is not a simple rectangular shape. In such cases, extra ventilation openings equivalent to up to 0.6% of the total plan area of the roof may be needed. Such ventilation openings would probably be provided as "mushroom cap"-type ventilators through the surface of the roof deck itself.

It should be noted that Approved Document F2 does not consider the use of vapour-control layers as an alternative to roof-space ventilation, unless a complete vapour barrier is installed. No mention is made of what would be necessary in order to demonstrate that such a barrier was in fact complete. In reality, the use of vapour checks is a standard part of cold "flat roof" construction, but is used in addition to ventilation of the roof space. Despite some pressure from the manufacturers of membrane materials, there seems to be a little confidence in the notion of using such a membrane in a more typical roof space to the exclusion of ventilation. This is

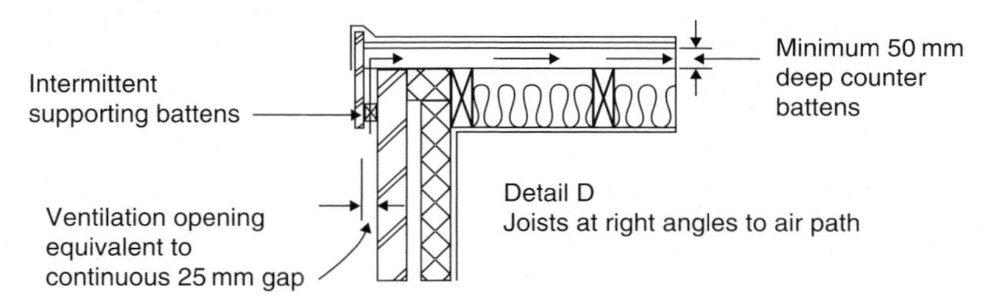

Figure 7.7 Roof ventilation less than 15°

due to the difficulties associated with ensuring that the membrane is perfectly vapour-tight at joints, and is also not damaged during construction.

Some difficulties may arise in cases where the roof void is not a continuous space. These may be encountered, for example in a case where a substantial extension has been added to a dwelling. Another possibility is where the roof space has been partitioned up to roof level so as to form a series of sub-compartments within the roof, as might sometimes be necessary to meet requirements for fire safety. In such cases, the best course of action, although not actually suggested by Approved Document F, is to treat each sub-compartment as a roof space in its own right, and hence provide appropriate ventilation provision so as to give Regulations the compliance in each space.

As in the case of ventilation of the occupied spaces in the dwelling, there is also an alternative means of compliance for roof-space ventilation. In this case, reference should be made to BS5250, Clauses 9.1, 9.2 and 9.4.

The use of the word *equivalent* with respect to ventilation openings should be stressed. The ventilation openings do not have to be continuous. The precise nature of the provision will depend on the detail of the eaves of the dwelling. If there is a soffit board, then the preference will usually be for a continuous opening. These can be bought as plastic mouldings (Figure 7.8) from a wide range of manufacturers. An integral

Figure 7.8 Trickle ventilator continuous strip

Figure 7.9 Twist and lock ventilator

insect screen is required within the ventilator, as indeed is the case with all other ventilation openings serving the roof. In some roof details, it may be necessary to pipe the air from the ventilator into the roof space. The use of pipes has an extra advantage in that it removes the risk of the eaves ventilation being reduced or even blocked out due to packing of insulation into the eaves. Positioning of insulation is critical. Packing reduces air flow; but on the other hand, cutting the insulation too short can lead to the formation of a cold bridge with consequent risk of surface and mould growth on the affected area of ceiling. The installation of continuous strip ventilators is very difficult if done as part of a refurbishment. In such circumstances, the preference would be for the use of the so-called "twist and lock" ventilators. A typical pattern of ventilator is shown in Figure 7.9. A hole of appropriate diameter is made in the soffit board by means of a cutting tool, and the ventilator is merely pushed into the hole and twisted to secure it in position. If there is no soffit board, then tile ventilators must be used at low level on the roof pitch itself, as shown in Figure 7.10. The design of these types of ventilators has improved in recent years, and a far greater range of products are available and these products are suitable for all types of roof coverings. The colour matchings are particularly good. Some ventilators are available for slate roofs which are so carefully designed so as to be almost invisible from the ground. This is an important consideration in conservation areas where standard patterns of ventilator might be deemed inappropriate on aesthetic grounds.

For pitched roofs, where the use of high-level ventilators is required, the first choice would be to use a ridge tile ventilator similar to those used as gas flue or PSV-system exhaust terminals, as shown in Figure 7.11. The provision needed would be of the order of 5000 mm^2/m of ridge length. Some propriety continuous opening systems are available, but do not seem to be used very often. In circumstances where ridge vents cannot be used, for example when the ridge board is not very long or where other vents have already been put in position (such as gas flue and PSV exhaust terminals), then it may not be physically possible to place another vent at ridge level, or else it might be considered that the ridge board may be weakened by doing so. In such circumstances, tile ventilators may be placed at high level on the roof pitches as an alternative.

More details about roof-space condensation and its control are given in Chapter 6.

Figure 7.10 Tile ventilator terminals

Figure 7.11 Ridge tile ventilator terminal

7.2.5 The effect of the 1995 Revisions to Approved Document F

In the past, there has been little effort to assess the effect of changes to the Building Regulations and their associated Approved Documents within a relatively short time from their date of introduction. It is pleasing to note that this has not proved to be the case with the 1995 Revisions to Approved Document F.

A recent piece of work,[15] carried out by University College London with the support of a consortium of interested firms and organisations with the support of the DETR (now DEFRA), investigated the influence of ventilation strategy on concentrations of a range of indoor air pollutants in the social housing, with a particular emphasis on carbon monoxide (CO). The study was carried out over a period of 2 years between 1996 and 1998, with the aim of collecting data during the winters of 1996–1997 and 1997–1998. A total of 45 occupied dwellings, all built since the changes to Approved Document F came into force, were monitored during the study: all of these were part of local authority, housing association or housing trust controlled. Despite the original intention to eliminate smoking as a variable within the sample group, in the end 39% of the dwellings had at least one smoker within the group of occupants. Within the sample, dwellings were equipped with a range of ventilation strategies conforming to the requirements of the 1995 Revision to Approved Document F. Each property was monitored for a period of 3 weeks during the winter season. Nitrogen dioxide was taken as an average value over the 3-week period, and house dust mite samples were taken once from two locations, usually the living room and a bedroom, on one occasion during the monitoring period. All dwellings were subjected to a fan pressurisation test in order to determine the envelope air leakage with all ventilation openings sealed.

The results showed several interesting findings. Detailed measurements in one dwelling showed that background CO concentrations were dependent on the external concentration, and that peak concentrations were associated with the use of the cooker. Consistently larger average and peak concentrations were found in kitchens. Peak CO concentrations at head height were higher than those at floor level. None of the properties was found to show a CO level in excess of World Health Organisation (WHO) guidelines. Measured peak values were between 10 and 30 parts per million (ppm), whilst the range of measured background levels was between 3 and 10 ppm. Houses with a smoker or smokers, exhibited CO levels which were on average 56% higher than those with no smoker amongst the occupants. The use of both *window openings* and *trickle ventilators* had effects on CO concentrations. In dwellings where windows were normally left open, it was found that mean CO levels were 32% higher than those where the windows were normally left closed. In contrast, mean CO levels in dwellings where trickle ventilators were normally kept closed were 37% higher than those where they were normally left open. Fan pressurisation test results indicated that the air-leakage rates for the dwellings were much lower than those monitored during the work carried out by the Building Establishment and its contractors in 1992.[16] The results showed no significant correlation between airtightness and type of ventilation strategy.

When asked about their perceptions of problems with condensation, 22% of the respondents said that in their opinion that it was occurring. Interestingly, this is very similar to the percentage of problem dwellings noted during the 1992 English House condition survey. No correlation was noted between measured air leakage and occupant perception of stuffiness or draughtiness. However, a weak relationship was noted between mean indoor air temperature and occupant perception of stuffiness.

7.2.6 *Part J*

The last revision to Approved Document J of the Building Regulations came into force in 2002. Part J is concerned with the safe installation of heat-producing appliance in the buildings. The scope of Part J is limited to fixed appliances that burn solid fuel, gas or oil. Incinerators come under the scope of Part J, but these have little if any significance to dwellings as sources of heat.

For the purposes of Approved Document J, heating appliances falling under its coverage are divided into two groups:

1. *Solid-fuel- and oil-burning appliances* with a rated heat output of up to 45 kW; these are referred to as Class A appliances.
2. *Gas burning appliances* with a rated heat output of up to 60 kW; these are referred to as Class B appliances.

Note: This is made for convenience within this chapter. It should be noted that this labelling is not used within the Approved Document.

Interestingly, no advice is given about the safe installation of appliances with higher-rated outputs than those specified above. The reason for this omission is probably that the maximum-rated outputs given are sufficient to cover the vast majority of heating requirements for dwellings. Thus the responsibility for the safe installation of larger appliances would be likely to fall to a qualified engineer as part of the overall building services package. The same cannot be said for the smaller heating appliances, and it is probably on this basis that the classifications within Approved Document have been made.

When comparing the two types of appliance, their differing characteristics should be taken into account. Gas-heating appliances (Class B) are carefully designed, and their manufacture is subject to quality assurance checks. Installation is also subject to substantial control and scrutiny. In contrast, the same may not be true for the Class A appliances. The other important characteristics of Class A appliances are that they have a far greater tendency to produce smoke and soot than Class B appliances if operating conditions are not ideal, and also generate far higher temperatures within their flues. Due to this, the requirements set out in Approved Document J for Class A appliances are far more exacting than those set out for Class B.

7.2.6.1 The supply of air

Regulation J1 states that:

> Heat-producing appliances are required to be provided with an adequate supply of air for the combustion of the fuel and for efficient operation of the chimney or flue.

In order to satisfy Regulation J1, there are two options:

- Firstly, the appliance may be of the room-sealed type, more often referred to as having a balanced flue.
- Secondly, alternative is that adequate ventilation should be provided in the room in which the appliance is situated.

The requirements for air supply depend on the class of appliance and the fuel type. Any open areas for ventilation purposes should be provided in addition to those recommended in Part F. There are three subgroups within Class A, and each has a different requirement. Ventilation for an *open solid-fuel appliance* should be provided by means of a permanent air inlet or inlets with a total free area equal to at least 50% of the throat opening of the appliance, the dimensions of which are to be determined by following BS8303 1986. For other *solid-fuel appliances*, the air inlet provision should be equivalent to at least 550 mm^2/kW of rated output over 5 kW. If such an appliance is fitted with a draught-stabiliser device, then overall ventilation will be reduced. To overcome this, extra inlet openings should be provided to a minimum of 300 mm^2/kW of rated output. The requirement for an *oil-burning appliance* is the same as for a solid-fuel appliance without a draught stabiliser.

Ventilation requirements for Class B appliances are dictated by appliance type. Any room in which a gas cooker is installed should have an openable window or other window to the outside air. The size of the opening window is not specified. If the room is less than 10 m^3 volume, then an extra permanent inlet provision of 5000 mm^2 must also be provided. Extra ventilation for rooms containing a room-sealed appliance is quite logically not required. Instead, Approved Document J requires that any Class B appliance located in a bathroom, shower room or private garage must be of the room-sealed type.

In the case of open-flued gas appliances, the requirement for permanent inlet area is 450 mm^2/kW of rated output over 7 kW.

Figures 7.12–7.16 give air supply requirements for a range of different heating appliances as stipulated by Approved Document J.

Type of appliance	Type and amount of ventilation[1]	
Open appliance, such as an open fire with no throat (e.g. a fire under a canopy)	Permanently open air vent(s) with a total free area of at least 50% of the cross-sectional area of the flue	
Open appliance, such as an open fire with a throat	Permanently open air vent(s) with a total free area of at least 50% of the throat opening area[2]	
Other appliance, such as a stove, cooker or boiler, with a flue draught stabiliser	Permanently open air vent(s) as below[3]:	Total free area
	First 5 kW of appliance rated output	300 mm^2/kW
	Balance of rated output	850 mm^2/kW
Other appliance, such as a stove, cooker or boiler, with no flue draught stabiliser	A permanently air entry opening or openings with a total free area of at least 550 mm^2/kW of appliance rated output above 5 kW	

Notes:

[1] Divide the area given in mm^2 by 100 to find the corresponding area in cm^2

[2] For simple open fires the requirement can be met with room ventilation areas as follows:

Nominal fire size (fireplace opening size) (mm)	500	450	400	350
Total free area of permanently open air vents (mm^2)	25,500	18,500	16,500	14,500

[3] Example: an appliance with a flue draught stabiliser and a rated output of 7 kW would require a free area of $(5 \times 300) + (2 \times 850) = 3200$ mm^2

Figure 7.12 Air supply (solid-fuel appliances)

Type of appliance	Type of ventilation
DFE fire in a fireplace recess with a throat	Air vent free area of at least 10,000 mm^2 (100 cm^2)
DFE fire in a fireplace with no throat (e.g. under a canopy)	Air vent free area sized as for a solid fuel fire (see Figure 7.12)
DFE fire with rating not exceeding 7 kW (net)	Permanently open air vents not necessary for appliances certified by a Notified Body as having a flue gas clearance rate (without spilling) not exceeding 70 m^3/h

Figure 7.13 Air supply (flued decorative effect (DFE) fires) (taken from Billington *et al.*[3])

7.2.6.2 Discharge of combustion products

Appropriate provision for the ingress of ventilation air must be matched by an appropriate capacity for the removal of combustion products. This will, of course, be achieved by the flue of the appliance. Regulation J2 states that:

> Heat-producing appliances are required to adequate provision for the discharge of the products of combustion to the outside air.

The principal objectives that Approved Document J seeks to achieve are that flues, flue pipes and chimneys are of adequate size; are either constructed

Location/type of appliance	Amount/type of ventilation
Appliance in a room or space	Ventilation direct to outside air
Open-flued	Permanently open vents of at least 500 mm²/kW (net) of rated input over 7 kW (net)
Room-sealed	No vents needed
Appliance in appliance compartment	Ventilation via adjoining room or space
Open-flued	From adjoining room or space to outside air: Permanently open vents of at least 500 mm²/kW (net) of rated input over 7 kW (net) Between adjoining room or space and appliance compartment, permanently open vents: at high level – 1000 mm²/kW input (net) at low level – 2000 mm²/kW input (net)
Room-sealed	Between adjoining space and appliance compartment, permanently open vents: both high and low levels – 1000 mm²/kW input (net)
Appliance in appliance compartment	Ventilation direct to outside air
Open-flued	Permanently open vents: at high level – 500 mm²/kW input (net) at low level – 1000 mm²/kW input (net)
Room-sealed	Permanently open vents: both high and low levels – 500 mm²/kW input (net)

Example calculation

An open-flued boiler with a rated input of 20 kW (net) is installed in a boiler room (appliance compartment, row 3 above) ventilated directly to the outside. The design of the boiler is such that it requires cooling air in these circumstances.
The cooling air will need to be exhausted via a high-level vent.
From above, the area of ventilation needed = 20 kW × 500 mm²/kW = 10,000 mm².
A low-level vent will need to be provided to allow cooling air to enter, as well as admitting the air needed for combustion and the safe operation of the flue.
From above the area of ventilation needed = 20 kW × 1000 mm²/kW = 20,000 mm².
These ventilation areas can be converted to cm² by dividing the results given in mm² by 100.
The calculated areas are the free areas of the vents (or equivalent free areas for proprietary ventilators).

Figure 7.14 Air supply (gas appliances other than decorative effect (DFE) fires or flueless types) (taken from Billington *et al.*[3])

or lined with suitable materials; discharge at roof level in a safe manner; and contain only such openings that are necessary for inspection, cleaning or efficient working. The construction of flues and material selection for liners is beyond the scope of this book. Readers seeking further information are referred to *The Building Regulations Explained and Illustrated* by Billington *et al.*[17]

Permitted discharge methods are again governed by appliance class. Class A appliances are allowed to discharge into balanced or low-level flues, factory-made or insulated chimneys, or flue pipes which discharge to the external air. For a balanced flue on an oil-burning appliance, no part

Type of flueless appliance	Maximum rated heat input of appliance	Volume of room, space or internal space[1] (m^3)	Free area of permanently open air vents (mm^2)
Cooker, oven, hob, grill or combination of these	Not applicable	Up to 5 5 to 10 over 10	10,000 5000* Permanently open vent not needed
Instantaneous water heater	11 kW (net)	5 to 10 10 to 20 over 20	10,000 5000 Permanently open vent not needed
Space heater[2]: (a) not in an internal space[3]	0.045 kW (net) per m^3 volume of room or space	All room sizes	10,000 **plus** 5500 per kW (net) >2.7 kW
(b) in an internal space[4]	0.09 kW (net) per m^3 volume of room or space	All room sizes	10,000 **plus** 2750 per kW (net) >5.4 kW

Notes:

[1] In the table above "internal space" means a space which communicates with several other room spaces (e.g. a hallway or landing).

[2] For LPG-fired space heaters which conform to BS EN 449: 1997 *Specification for dedicated liquified petroleum gas appliances. Domestic flueless space heaters (including diffusive catalytic combustion heaters)*, follow the guidance in BS5440: Part 2: 2000.

* If the room or space has a door direct to outside air, no permanently open air vent is needed.

[3] A space heater is to be installed in a lounge measuring $5 \times 4 \times 2.5 = 50\,m^3$. Its maximum rated input should not exceed $50 \times 0.045 = 2.25$ kW (net).

[4] A space heater installed in a hallway to provide background heating has a rated input of 6 kW (net). It will need to he provided with $10{,}000 + 2750 \times (6 - 5.4) = 11{,}650\,mm^2$ of permanently open ventilation.

Figure 7.15 Air supply (flueless gas appliances) (taken from Billington *et al.*[3])

of the flue terminal must lie within 600 mm of any opening to the building. This is in order to minimise the risk of combustion products being re-entrained in ventilation air flowing back into the building. The flue must be guarded against any blockage. Flue sizes should be in accordance with those given in Figures 7.17 and 7.18, and should in any case never be less than the size of the discharge outlet on the appliance itself.

The only types of permitted openings into the flue are: an opening for inspection and cleaning, which must have a non-combustible double-cased gas-tight cover; and an opening for a draught diverter, draught stabiliser or explosion door, which must be made of a non-combustible material.

No flue must communicate with more than one room or internal space within a building. However, an inspection or cleaning opening may be located in a room other than the one in which the served appliance is located. If desired, a single flue may also serve more than one heating appliance in the same room.

Location/type of appliance	Amount/type of ventilation
Appliance in a room or space	Ventilation direct to outside air
Open-flued	Permanently open vents of at least 550 mm²/kW of rated output over 5 kW[1]
Room-sealed	No vents needed
Appliance in appliance compartment	Ventilation via adjoining room or space
Open-flued	From adjoining room or space to outside air: Permanently open vents of at least 550 mm²/kW of rated output over 5 kW Between adjoining room or space and appliance compartment, permanently open vents: at high level – 1100 mm²/kW output at low level – 1650 mm²/kW output
Room-sealed	Between adjoining space and appliance compartment, permanently open vents: both high and low levels – 1100 mm²/kW output
Appliance in appliance compartment	Ventilation direct to outside air
Open-flued	Permanently open vents: at high level – 550 mm²/kW output at low level – 1100 mm²/kW output
Room-sealed	Permanently open vents: both high and low levels – 550 mm²/kW output

Notes:

[1] Increase the area of permanent ventilation by a further 500 mm²/kW output if appliance fitted with draught break.

Example calculation

An open-flued boiler with a rated output of 15 kW is installed in a cupboard (appliance compartment, row 2 above) ventilated via an adjacent room. Since the boiler output exceeds 5 kW, permanent ventilation openings will be needed in the adjacent room in addition to the vents between the cupboard and the room disigned to provide combustion and cooling air.
Area of permanent vents to outside air needed in adjacent room = (15 kW − 5 kW) × 550 mm²/kW = 5500 mm².
The cooling air will need to be exhausted via a high-level vent.
From above, the area of ventilation needed = 15 kW × 1100 mm²/kW = 16,500 mm².
A low-level vent will need to be provided to allow cooling air to enter, as well as admitting the air needed for combustion and the safe operation of the flue.
From above, the area of ventilation needed = 15 kW × 1650 mm²/kW = 24,750 mm²
These ventilation areas can be converted to cm² by dividing the results given above in mm² by 100.
The calculated areas are the free areas of the vents (or equivalent free areas for proprietary ventilators).

Figure 7.16 Air supply (oil appliances) (taken from Billington *et al.*[3])

Great care must be taken to ensure that exhaust-flue gases are not drawn back into the dwelling. Several restrictions are given in Approved Document J Section 2, which seek to prevent this happening. The discharge outlet must be at least 1 m above the top of any openable part of a skylight or window. The same restriction applies to any ventilator or similar opening which is located on a roof or external wall, and is not greater than 2.3 m horizontally from the top of the discharge flue or chimney, as shown in Figure 7.19. The discharge must also be 600 mm above the top of any part of any adjoining building which is not greater than 2.3 m horizontally from the top of the flue or chimney.

Installation[1]	Minimum flue size
Fireplace with an opening of up to 500 mm × 550 mm	200 mm diameter or rectangular/square flues having the same cross-sectional area and a minimum dimension not <175 mm
Fireplace with an opening in excess of 500 mm × 550 mm or a fireplace exposed on two or more sides	If rectangular/square flues are used the minimum dimension should not be <200 mm
Closed appliance of up to 20 kw rated output which: (a) burns smokeless or low volatiles fuel[2] (or) (b) is an appliance which meets the requirements of the Clean Air Act when burning an appropriate bitminous coal[3]	125 mm diameter or rectangular/square flues having the same cross-sectional area and a minimum dimension not <100 mm for straight flues or 125 mm for flues with bends or offsets
Other closed appliance of up to 30 kW rated output burning any fuel	150 mm diameter or rectangular/square flues having the same cross-sectional area and a minimum dimension not <125 mm
Closed appliance of above 30 kW and up to 50 kW rated output burning any fuel	175 mm diameter or rectangular/square flues having the same cross-sectional area and a minimum dimension not <150 mm

Notes:

[1] Closed appliances include cookers, stoves, room heaters and boilers
[2] Fuels such as bituminous coal, untreated wood or compressed paper are not smokeless or low volatiles fuels.
[3] These appliances are known as "exempted fireplaces"

Figure 7.17 Flue sizes in chimneys (from Approved Document J) (taken from Billington *et al.*[3])

Intended installation	Minimum flue size	
Radiant/convector gas fire	New flue: Circular Rectangular	 125 mm diameter 16,500 mm^2 cross-sectional area with a minimum dimension of 90 mm
	Existing flue: Circular Rectangular	 125 mm diameter 12,000 mm^2 cross-sectional area with a minimum dimension of 63 mm
Inset line fuel effects or DFE fire within a fireplace opening up to 500 mm × 550 mm	Circular or Rectangular	Minimum flue dimension of 175 mm
DFE fire installed in a fireplace with an opening in excess of 500 mm × 550 mm	Calculate in accordance with xxxxxxxx and the notes referring to similar sized openings for appliances burning solid fuel in section xxxxx	

Figure 7.18 Flue sizes for gas-fired appliances (taken from Approved Document J) (taken from Billington *et al.*[3])

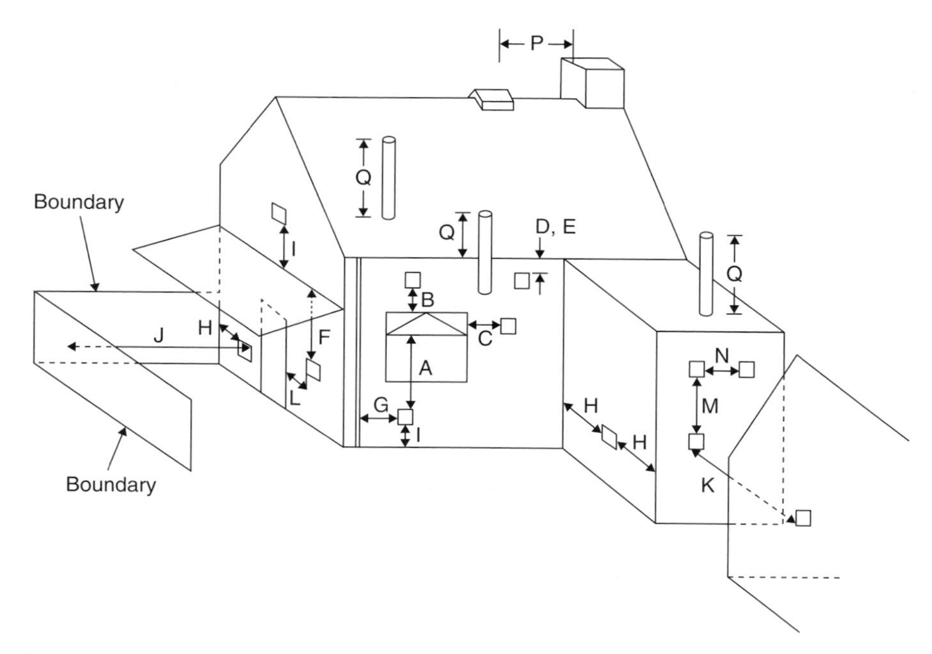

Figure 7.19 Locations of outlets from flues serving gas appliances (taken from Approved Document J). For labelling details, refer to Figure 7.20

The discharge height of any flue pipe or flue in a chimney is also governed by the pitch of the roof above which discharge is taking place. The discharge should be at least 1 m above the highest point of contact. In addition, for roof pitches in excess of 10°, the discharge should be at least 2.3 m (measured horizontally) from the roof ridge, as shown in Figure 7.20.

If the flue pipe or chimney passes through the roof within 2.3 m of the ridge line and the pitch of both slopes of the roof is 10°, the top of the flue pipe or chimney must not be less than 600 mm above the ridge line.

The only exception to these restrictions is that of the oil-fired pressure jet-heating appliance (likely to be uncommon in dwellings). With such appliances, the flue discharge may be located anywhere above the ridge-line of the building.

Other specific restrictions for the design of flue systems are set out. These are intended to avoid possible reductions in air flows by increasing the resistance of the flue. For example, it is recommended that horizontal runs should be avoided in flue systems. The exceptions to this are balanced-flue systems, back outlets from solid-fuel-burning appliances and low-level flues for oil-burning appliances. Similarly, bends in flue systems are also discouraged. If a situation arises where a bend is unavoidable, it should not be greater than 45° to the vertical for an oil-burning appliance or 30° for a solid-fuel-burning appliance.

Minimum separation distances for terminals (mm)					
Location		Balanced flue		Open flue	
		Natural draught	Fanned draught	Natural draught*	Fanned draught
A	Below an opening[1]	Appliance rated heat input (net) 0–7 kW 300 >7–14 kW 600 >14–32 kW 1500 >32 kW 2000	300		300
B	Above an opening[1]	0–32 kW 300 >32 kW 600	300		300
C	Horizontally to an opening[1]	0–7 kW 300 >7–14 kW 400 >14 kW 600	300		300
D	Below gutters, soil pipes or drain pipes	300	75		75
E	Below eaves	300	200		200
F	Below balcony or car port roof	600	200		200
G	From a vertical drain pipe or soil pipe	300	150†		150
H	From an internal or external corner or to a boundary alongside the terminal[2]	600	300		200
I	Above ground, roof or balcony level	300	300		300
J	From a surface or a boundary facing the terminal[2]	600	600		600
K	From a terminal facing the terminal	600	1200		1200
L	From an opening in the car port into the building	1200	1200		1200
M	Vertically from a terminal on the same wall	1200	1500		1500
N	Horizontally from a terminal on the same wall	300	300		300
P	From a structure on the roof	Not applicable	Not applicable	1500 mm if a ridge terminal. For any other terminal, as given in BS5440-1:2000	Not applicable
Q	Above the highest point of intersection with the roof	Not applicable	Site in accordance with manufacturer's instructions	Site in accordance with BS5440-1:2000	150

Notes:

* For locations A–N natural draught should not be used.
† This dimension may be reduced to 75 mm for appliances of up to 5 kW input (net).
[1] An opening here means an openable element, such as an openable window, or a fixed opening such as an air vent. However in addition, the outlet should not be nearer than 150 mm (fanned draught) or 300 mm (natural draught) to an opening into the building fabric formed for the purpose of accommodating a built in element, such as a window frame.
[2] Boundary. Smaller separations to the boundary may be acceptable for appliances that have been shown to operate safely with such separations from surfaces adjacent to or opposite the flue outlet.

Figure 7.20 Locations of outlets from flues serving gas appliances (taken from Approved Document J)

When choosing the locations of inspection and cleaning points for a Class A system, it should be borne in mind that it is likely that debris will be collected within the flue. If the flue does not rise vertically from the appliance, then a debris collection space should be incorporated into the inspection and cleaning provision.

The requirements for the discharge of Class B appliances differ slightly from those for Class A. Four subgroups of appliance are identified:

1. cooking appliances;
2. balanced-flue appliances;
3. decorative log or solid-fuel fire-effect appliances;
4. individual, natural draught open-flued appliances, such as boilers and back boilers.

In the case of balanced-flue appliances, the same requirements as per Class A apply. The exception is that in a case where an appliance terminal is situated wholly or partly below an opening, such as a window or ventilator, no part of the flue terminal should be within 300 mm vertically to the bottom of the opening. For a flue serving a gas fire, the cross-sectional area should be at least 12,000 mm^2 for round flues and 16,500 mm^2 for rectangular flues. This reflects the greater resistance to air flow within a rectangular duct compared to a round duct of the same cross-sectional area. For any other appliance, the cross-sectional area of the duct should not be less than the area of the outlet of the appliance in question.

Other parts of the Approved Document show small differences from the requirements for Class A appliances. In the case of permitted openings to flues, it is stated that an opening for cleaning and inspection plus a draught diverter, stabiliser or explosion door are permitted; in the case of Class B appliances, the wording is such that either a cleaning and inspection opening *or* one of the other types of opening is allowed. For Class B appliances, it is stipulated that a flue terminal must be fitted if any dimension across the flue outlet is less than 175 mm. The terminal must be located so that air may flow freely over it at all the time, and no part of the terminal must be within 600 mm of any opening into the building itself. No such requirement for the use of a terminal exists for Class A appliances. Any bends in flues should not be greater than 45° from the vertical.

7.3 Building Regulations in Northern Ireland

The Building, Northern Ireland Regulations,[18] hereafter referred to as the Northern Ireland Regulations, are organised rather differently to the Regulations for England and Wales. Instead of the system of Approved Documents and generalised regulations, the Northern Ireland Regulations rely on self-contained regulations (the Technical Booklets) which are presented in some detail, together with explanatory notes which do not form part of the Regulations themselves and are quite clearly labelled as not doing so. Reference to other documentation is very sparse within the Northern Ireland Regulations. The labelling of the different Regulations is not coincident with the system used within the English and Wales Regulations. One particular point of note is that Northern Ireland Regulations may be changed by means of the issue of an amendment booklet (the actual Technical Booklets may not be revised in their entirety).

Part C of the Northern Ireland Regulations deals with site preparation and resistance to moisture. There is no specific mention of the control of the ingress of landfill gases: instead, this is covered by the very general requirement of Regulation C2(1). Regulation C3(1) deals specifically with radon ingress. The action level is set at the 200 Bq/m^3 used in England and Wales. Within Regulation C4, Northern Ireland is divided into four zones:

- The first zone is not assigned a designation: in this zone, radon is not deemed to pose a risk, and therefore no measures have to be taken.
- The other three zones are designated by letters which indicate the level of radon protection that is required.
- In Zone A, there is deemed to be a probability between 1% and 3% of the radon concentration exceeding 200 Bq/m^3, whilst in Zones B and C the probabilities are between 3% and 10%, and over 10%, respectively.

Radon measures are specified according to Figure 7.21 for the case of any new dwelling, or else in circumstances where a dwelling not already incorporating radon control measures is being extended by greater than 30 m^2 area. In cases where a dwelling is already incorporating radon control measures, which is being either extended or altered, it is stipulated that the same control measures are maintained or extended as appropriate throughout the dwelling in question.

Zone designation	Where the floor next to the ground is above a void or where there is a void behind the wall	In any other case
A	The void shall be adequately ventilated	A radon proof membrane shall be incorporated into the construction
B	The void shall be adequately ventilated and a radon proof membrane shall be incorporated into the construction	A radon proof membrane shall be incorporated into the construction and a sump and stub duct shall be installed
C	The void shall be adequately ventilated and a radon proof membrane shall be incorporated into the construction	A radon proof barrier shall be incorporated into the construction and a sump and stub duct shall be installed

Figure 7.21 Radon compliance requirements for Northern Ireland (taken from Part C)

Ventilation requirements are covered within *Technical Book Part K of the Building, Northern Ireland Regulations*. The definitions of types of rooms are very similar to those used in the England and Wales Building Regulations, although it should be noted that some of the definitions are given within other parts. No mention is made of utility rooms. The main Regulation (Part K2) states that:

> Adequate means of ventilation shall be provided for people using any habitable room, kitchen, bathroom, common space or sanitary accommodation in any building.

Part K3 contains what are called "deemed to satisfy" provisions for ventilation. At the start of Part K3, definitions are given of a wide range of terms related to ventilation which are not all given in the England and Wales Regulations. Most noticeable amongst these is "top of the wall". It is fair to say that Part K of the Building, Northern Ireland Regulations is not written very clearly. Careful study is required in order to understand fully.

Table 2.1 of Approved Document K, presented here as Figure 7.22, sets out requirements for ventilation directly to external air required for specified rooms in order that the requirements are satisfied. The requirements are similar to those for England and Wales. The seven notes are of some importance and should be read carefully in conjunction with the table (Figure 7.22).

There are important conditions placed on the use of extract ventilation within a room that contains an open-flued solid-fuel-burning appliance. In such circumstances, mechanical extract ventilation is specifically proscribed. There is no compulsion to use either PSV or mechanical ventilation in a room with an open-flued appliance which has a flue with a free area equivalent to at least a 125 mm duct and also if the combustion-air and dilution-air inlets are permanently open, even when the appliance is not in use. It is presumed that the latter clause applies only to non-solid-fuel-burning appliances; there seems to be a lack of clarity in Note 4 (Figure 7.22) at this point.

7.4 Scotland

In Scotland, the system of building control is similar to that of England and Wales. The basis for the system is contained within the Building Standards (Scotland) Regulations; hereafter referred to as the Scottish Regulations. It is interesting to note that in contrast to the situation in England and Wales, the word "standards" finds its way into the title of the Scottish Legislation. At present, the Scottish Regulations are made by the Secretary of State for Scotland under powers given by the Building (Scotland) Act 1959. It is envisaged that with the major constitutional changes imminent in Scotland, at least some of the powers governing the Scottish Regulations will be devolved to the new Scottish Parliament or one of its agencies.

Room[1]	**Rapid ventilation opening(s)** (minimum free area)	**Background ventilation opening(s)[2]** (minimum free area, mm^2)	**Mechanical exact ventilation[3,4]** (nominal air-flow rates)
Habitable room	1/20th of floor area	8000	
Kitchen[5]	1/20th of floor area	4000	30 l/s adjacent to a hob[6] or 60 l/s elsewhere
Utility room	1/20th of floor area	4000	30 l/s
Bathroom (with or without WC)	1/20th of floor area	4000	15 l/s
Sanitary accommodation (separate from bathroom)	1/20th of floor area[7]	4000	

Notes:

[1] Where a room serves a combined function such as kitchen-diner, the individual provisions for rapid, background and mechanical extract ventilation need not be duplicated provided that the greater or greatest provision for the individual functions in the above table is made.

[2] As an alternative to the background ventilation provisions listed in the above table, background ventilation openings equivalent to an average of 6000 mm^2 per room may be provided but no room shall have a background ventilation opening of >4000 mm^2.

[3] As an alternative to mechanical extract ventilation, PSV may be provided. Where PSV is provided if shall be designed and constructed in accordance with BRE IP 13/94 or a valid BBA Certificate.

[4] Mechanical extract ventilation shall not be provided in a room where there is an open-flued solid fuel burning appliance. Mechanical extract ventilation (or PSV) need not be provided in a room with an open-flued appliance which has a flue having a free area at least equivalent to a 125-mm diameter duct and the appliance's combustion air inlet and dilution air inlet are permanently open when the appliance is not in use.

[5] This provision is for a domestic size kitchen where the appliances and usage are of a domestic nature. Guidance on the ventilation required for commercial kitchens is given in CIBSE Guide B, Tables B2.3 and B2.11

[6] Adjacent to a hob means either:
(a) incorporated within a cooker hood located over the hob; or
(b) located near the ceiling within 300 mm of the centreline of the space for the hob.

[7] As an alternative, mechanical extract ventilation at 6 l/s may be provided.

Figure 7.22 Ventilation requirements for Northern Ireland (taken from Part K)

The current Scottish Regulations are contained within the document Technical Standards for Compliance with the Building Standards (Scotland) Regulations.[19] The document is substantial, consisting of nearly 500 A4 pages. The sections are now individually available; it is a wise move in the days of web-downloadable documents. Not only does the Technical Standards document contain the actual regulatory requirements themselves, but it also explains key points and, where appropriate, gives examples of calulations that might have to be carried out for purposes of demonstrating compliance. However, it is not self-contained; a substantial list of publications referred to within its main text are presetned as an Appendix.

The Technical Standards document consists of 17 parts, labelled as Parts A to T inclusive. From the ventilation point of view, the parts of interest with regard to domestic ventilation are Part D (Structural Fire Precautions), Part F (Heat Producing Installations and Storage of Liquid and Gaseous Fuels), Part G (Preparation of Sites and Resistance to Moisture) and Part K (Ventilation of Buildings).

Part D contains requirements for fire resistance of ventilation ductwork. To achieve compliance, the ductwork must either meet the requirements of BS5588 Part 9[20] or else be provided with automatic dampers, shutters or other sealing device at any point where the duct passes through a structural element, so that in the event of fire the required fire resistance of the structural element is maintained and the flow of hot gases and air is prevented. *It is interesting to note that the word* smoke *is not used within this part of the Regulations.* An automatic duct-closure device is not required in cases where a cavity barrier is being penetrated by ductwork which is of at least 1.2-mm wall thickness and is also continuous throughout the cavity on both sides of the cavity barrier. Alternatively, the duct could either be of a construction or else within a construction which has at least the fire resistance of the structural element which is being penetrated. In this case, automatic-closure devices would only be required at points where either the duct or its enclosure do not have the same fire resistance as the structural element. A requirement for the automatic-closure device to prevent the passage of hot gases or air is not stipulated.

If the duct passes through a fire compartment wall or floor, and has an opening or openings to one compartment within the building in question, then compliance will be achieved if the ductwork has at least the fire resistance of the compartment wall or floor through which it passes. For buildings of purpose group 1 (dwellings), a ventilation duct serving only sanitary facilities and passing through a separating floor must have at least the fire resistance required for a separating floor, and any branches into the duct must be shunt ducts which enter the main duct not less than 900 mm above their inlets. This requirement covers all the dwellings of all heights, which is in contrast to the England and Wales Approved Document. Finally, all ducts must be provided with extract grilles fitted with non-return shutters.

Part F contains requirements for fixed heat-producing appliances burning solid, liquid or gaseous fuels.

Any appliances must be so constructed and installed that it operates safely, that the products of combustion are not a hazard to health and that it receives enough air for its safe operation.

The requirements for the removal of combustion products and the supply of fresh air are very similar to those given within Approved Document J of the England and Wales Regulations. One minor difference is that the size ranges of hearths do not match up, as indeed to all appliance sizes and flue diameters. Within the Scottish Regulations, the dimensions for permissible rectangular ducts are given, rather than merely stating that the area should be equivalent to that of the prescribed circular duct for a given application.

Part G of the Scottish Regulations corresponds in the most part to Part C of the England and Wales Regulations. However, there are signficant differences. The major thrust of the Scottish Regulations is the moisture aspects site preparation and protection. There is a section on the prevention of condensation, both surface and interstitial, that is not present within the England and Wales Regulations. Requirements G4.1 and G4.2 require that all building elements must minimise the risk of both surface and interstitial condensation where these might damage the health of people. Requirement 2.6 states that moisture must be prevented from entering the building from the sub-site. The requirements of this regulation would be deemed to be satisfied if the guidance within BS5250 1989 were to be followed.

Reference to roofs is made. Once again, compliance can be achieved by following the guidance in BS5250 1989. However, in the introduction to Part G, specific concerns about the use of cold deck roofs in Scotland is expressed, even in cases where the roof is properly ventilated.

Approved Document F of the England and Wales Regulations contains requirements for the ventilation of sub-floor voids with respect to the control of moisture ingress. In the Scottish Regulations, the relevant requirements are separated from the overall ventilation requirements and are instead to be found within Part F. Different requirements are set out for suspended timber and concrete floors. For suspended timber floors, permanent ventilation of the under-floor space is to be provided in two external walls on opposite sides of the building at a rate of either 1500 mm^2 for at least every metre run of the wall, or 500 mm^2 for at least every square metre. The latter method of compliance is not mentioned within Part C4 of the England and Wales Regulations. The use of the word "least" implies that the stated requirements are minima. In addition, the Technical Standard states that the open area should also be provided in internal sleeper walls or similar internal obstructions, in order that ventilation be maintained in all the parts of the sub-floor.

In common with Part C of the England and Wales Regulations, Part G of the Scottish Regulations also covers the issue of contaminated sites. However, the approach within the Scottish Regulations is rather different. There is a clear requirement (Regulation 16.1) that sites and ground adjacent to sites should be prepared and treated so as to protect the building and its occupants from harmful or dangerous substances, matter in the surface soil and vegetable matter. Rather than making legal requirements for compliance, the reader is required to refer to an Appendix to Part G. This is clearly labelled as neither forming part of the technical standards nor representing any deemed to satisfying any requirements. Rather, the Appendix is afforded the status of outline guidance. In terms of ventilation aspects, the only relevant information is contained within Table 2 of the Appendix, where it is recommended that any building constructed on a site which is possibly contaminated by flammable, explosive and asphyxiating gases, including methane and carbon dioxide, should be free from unventilated voids. This piece of outline guidance could be interpreted as meaning that this measure would be desirable when building on a site which may be contaminated by landfill gas. No guidance is provided as to what constitutes an adequate level of ventilation. This in itself is a major deviation from the requirements in Part C of the England and Wales recommendations. However, the most striking difference between the two pieces of legislation is the total omission of any reference to the control of radon ingress. This is most surprising, bearing in mind that granite is quite common in Scotland, and therefore high concentrations of radon might reasonably be expected to be present.

The stated objective of Part K of the Scottish Regulations is as follows:

> To ensure reasonable provision for an adequate supply of air for human occupation of a building.

Whereas compliance with Approved Document F for England and Wales now ensures that all requirements of the Factories Act are met, premises to which the Factories Act apply are specifically excluded from the requirements of Part K of the Scottish Regulations. Within the introduction to the Scottish Part K, it is acknowledged that the energy consumption of a building can be significantly affected by the choice of ventilation strategy, and it is stated that "a thorough assessment of natural as against mechanical ventilation should be made". In contrast to the provisions within Approved Document L of the England and Wales Regulations, this statement would appear to be targeted at all buildings falling under the remit of the Scottish Regulations, including the dwellings. The introduction acknowledges that improved standards of airtightness lead to increased risk of condensation in building, and particularly so in dwellings. The ventilation rates given for moisture-producing areas are deemed to be the minimum necessary for "combatting" condensation. The choice of word may be taken as being significant.

Regulation 23 states that:

> A building to which this regulation applies shall have means of providing an adequate supply of air for users of the building.

In Section K1, the volume of any space within a building is defined as the internal cubic capacity of the space, with two exceptions. In the case of a space used for the parking of vehicles, any volume within the space greater than 3 m above floor level is to be disregarded. The same applies to any volume in other spaces greater than 6 m above floor level. Section K2 states that:

> A building other than a garage must have adequate provision for ventilation by natural means, mechanical means, or a combination of natural and mechanical means.

Four exceptions to this are given. Rooms for controlled temperature storage need not be ventilated. Similarly, there is no requirement to ventilate a room with a floor area of less than $4\,m^2$. Any room where the space allowance per occupant is $3\,m^3$ or less must be mechanically ventilated. Finally, the use of passive stack systems is allowed in buildings of purpose group 1, that is, houses, flats and maisonettes. Furthermore, if such systems are installed, no storey of the building must be at a height of greater than 11 m. The effect of this latter clause places a restriction on the use of PSV in blocks of flats in Scotland.

Regulation K4 sets out the general ventilation requirements. Ventilation must be made to the outside air, except that a ventilator serving a room in a dwelling may discharge into a conservatory, provided that the ventilation of the conservatory itself is to the outside air. The ventilation provision for the conservatory must be calculated on the basis of the floor areas of both the conservatory and for the room in question.

7.5 British Standards

When considering the role of British Standards, it is important to understand the nature of the British Standards Institution (BSI), which exists in order to produce standards over a wide range of subjects. The number of British Standards runs into tens of thousands, and the list continues to grow every year. BSI earns its income primarily by means of subscription and by sales of publications. These publications are not cheap, and BSI has hitherto protected its copyright with great care. Within the 2 years, several electronic database(e.g. Barbour) have started to carry selected British Standards. Despite a stipulation that any copies taken must be destroyed within 1 month, BSI is probably in the same position of any other organisation or individual making publications available in electronic format. It will be very interesting to see the effect on its income.

The BSI organisation itself is relatively small. It relies on the contributions of volunteers; often seconded by their companies, trade associations, professional organisations, government departments and agencies. The financial pressures of recent years have made it extremely difficult to sustain the activities of BSI. The involvement of commercial interests brings the inevitable risk that the documents produced by BSI might be viewed by some as being tainted by vested interest. British Standards represents the attempts of committees to produce documents acceptable to the majority of members. They were never intended to be sources of new ideas. Despite such reservations, British Standards continue to be oft-quoted sources of information.

Another important function of the BSI is the management of the UK involvement with CEN standards and Eurocodes. BSI nominates National Representatives to all CEN and ISO committees.

Only a relatively small number of British Standards relate to buildings. Of these, an even smaller number have any significance in the area of domestic ventilation. The most important of these are discussed below.

7.5.1 *BS5250*

This standard is entitled *Code of Practice for the Control of Condensation in Buildings*. It was originally issued in 1985, and a revised edition was issued in 1989.[8] It is understood by the author at the time of writing that there are no current plans to update the standard. The standard is very wide reaching, and is not confined to issues of domestic ventilation. Rather, it seeks to present itself as a holistic approach to the control of condensation. It stresses the importance of the threefold approach to surface condensation, namely the *provision* of adequate ventilation, the *maintenance* of adequate internal temperatures and adequate insulation. The causes and effects of condensation are well summarised within the Standard. Table 1 of Appendix B is a regularly quoted source of data about the production of moisture as a result of occupancy (see Figure 2.2). Appendix C discusses amongst the effect of excessive ventilation on relative humidity. Examples are given of good detailing for the avoidance of cold bridging which might lead to surface condensation in an otherwise soundly designed dwelling.

The Standard gives a wide range of recommendations for the ventilation of various areas within the dwelling. It is interesting to note that PSV was mentioned in BS5250 a long time before Approved Document F. The ventilation of roof spaces is given particular consideration. This is quite understandable, since this was a very contentious issue when BS5250 was first issued. Although not strictly relevant to this text, BS5250 also discusses interstitial condensation calculations in great detail.

Conformation to BS5250 is quoted as a possible means of compliance within Approved Document F. It is not surprising that there are several discrepancies between the two documents, given the time elapsed since the last revision of the Standard. For example, the recommendations for the provision of trickle ventilation within BS5250 are at variance with those in the 1995 version of Approved Document F. The advice given about the use of PSV within BS5250 is vague in the extreme, and attempts to demonstrate conformity with its recommendations as an alternative means of compliance with the requirements of Approved Document F have thus far met with very little success.

In summary, the main text of BS5250 is still a document of some interest. However, the need to revise the document has been recognised, and a new version came into circulation in 2002.[9] The approach of the new document is somewhat different to the old one, as might be expected. In terms of ventilation provision, a conscious effort has been made to tie the new version not only to the current Approved Document F for England and Wales, but also to the requirements for Scotland and Northern Ireland. The energy implications and costs of ventilation are discussed, and the concept of risk assessments in accordance with the procedures outlined in current EN and ISO standards is introduced. In this context, it is very interesting to note that Appendix D.2 is highly critical of the Glaser method for the determination of interstitial condensation risk, and seeks to push users to more suitable (but unfortunately much more complex) prediction tools. The material presented concerned with design principles is very much more detailed than in the 1989 version.

7.5.2 *BS5925*

This Standard is entitled *Code of Practice for Ventilation Principles and Designing for Natural Ventilation*. It was originally issued in 1980 and its second edition was issued in 1991.[14,21] No revisions have been made to it since then. BS5925 is an interesting collection of information about the need for ventilation, how to calculate air change rates and how to make adequate provision. The scope of the Standard is not confined to dwellings.

BS5925 is presented in the usual thorough manner expected from BSI publications. It contains information about requirements for ventilation within the buildings with respect to controlling a range of pollutants. One appendix (Appendix A) is dedicated to the issue of thermal comfort of building occupants; although interestingly, very little mention is made of the need for indoor air temperature control by means of ventilation. One section of the Standard is devoted to a discussion of the factors that might influence the choice between natural and mechanical ventilation in a particular situation. The document, as might be expected for a BS, does not have any kind of evangelical pitch. Indeed, it is difficult to avoid noticing that in several parts of the Standard, the tendency is for the reader to be rather discouraged from considering the use of natural ventilation. The limitations of natural ventilation are stressed, whilst little mention is made of its benefits. Instead, some fairly precise advice is given about circumstances where the use of mechanical ventilation is either an absolute necessity or else desirable. A particularly questionable example of a situation considered by the Standard to be one where mechanical ventilation is desirable is "dwellings, in order to remove odours and excessive moisture from bathrooms and kitchens".

Section 3 of the Standard discusses the range of physical parameters affecting natural ventilation, together with means of determining them. These are dealt with in much the same way as Chapter 3. Sample data is included for the use of the reader; for example, a collection of mean daily temperatures for 12 meteorological stations in the UK for the period 1941–1970.

Of some considerable use are the equations given in Section 3 of the Standard for the estimation of air-flow rates due to natural ventilation, both for simple buildings and for single-sided ventilation as might be more commonly encountered in offices rather than dwellings. These are described in more detail in Chapter 3 of this book.

7.6 European aspects

The issue of European harmonisation is a vexed one from the political perspective: fortunately this is none of our concern here. From the point of view of regulatory implications for construction in general and ventilation in particular, some consideration of the European dimension is important.

One of the key objectives of the European Community (EC) is to eliminate technical barriers to trade between its member states. The sheer size of the construction sector means that it was bound to be subject to significant legislation. This came in the form of the Construction Products Directive (CPD).

Essentially, the CPD is one of a series of New Approach Directives. These cover a wide range of products, and are concerned mainly with health and safety matters. Compliance with Directives is through a series of European Technical specifications, which are developed by the European Standards organisations (CENELEC and CEN), with the involvement of the European Organisation for Technical Approvals. The main advantage of the New Approach Directives system is that it allows basic legislative frameworks to be agreed on without being bogged down in detailed negotiations/arguments about actual standards and associated technical approvals for particular aspects of the frameworks.

New Approach Directives list essential requirements for particular aspects of regulation. In the case of the CPD, six essential requirements are given. These are titled as:

- Mechanical resistance and stability.
- Safety in case of fire.
- Hygiene, health and the environment.
- Safety in use.
- Protection against noise.
- Energy economy and heat retention.

For each essential requirement, an interpretative document exists. The purpose of an interpretative document is to provide linkage to the actual specific requirements for product performance. An interpretative document includes all requirements that have to be met by products so as to be deemed to comply with regulations in all European Union (EU) member states. In other words, the requirements contained within an interpretative document are mandatory. To further complicate matters, there may be other non-mandatory items of specifications which may be adhered to on a non-voluntary basis. These will be contained within various standards documents other than the relevant interpretative document. The standards will in effect be the European Standards (the so-called Euro Norms or Ens) relevant to the product in question. It is intended that Ens should ideally be confined to statements of performance requirements, rather than becoming involved in setting out prescriptive requirements for the manufacture or composition of the product. One reason for this approach is that it enables standards to be formulated without causing conflicts with different construction practices within the member states. Any standards document must make very clear which requirements are mandatory in order to satisfy the CPD (the harmonised standard) and those that are not (the non-harmonised or voluntary standard).

The European Commission has in effect prioritised the production of standards by dividing them into three groups:

- Group A standards are concerned with design, construction and execution.
- Group B standards are linked to product characteristics.
- Group Bh contains the so-called horizontal standards of test methods for a range of properties that are intended to be applied across all interpretative documents.

The Commissions priority has been to deal with Group B standards first. This has been done so that the CE product-marking scheme could be introduced as quickly as possible. Group A standards are the responsibility of the member states as they must incorporate them into their national regulations. One effect of this approach, and one that is not widely appreciated, is that any changes to Building Regulations for any part of Great Britain must be approved by the Commission before they can pass on to the Statute Book.

A mandate is in effect as a contract between standards making bodies (e.g. CEN and CENELEC) and the EC for the production of the appropriate harmonised standards in support of the CPD. A standard for a particular product group is deemed to be harmonised once the final mandate has been issued by the EC, and reference to it has been published in the *Official Journal of the European Community*. Thirty-three draft mandates have been issued to CEN and CENELEC with respect to the use products in terms of construction elements. Each mandate contains a list of all products associated with a particular intended use.

The production of Ens is the responsibility of a large number of Technical Committees (TCs). Of these, the one that is of direct relevance to domestic ventilation is TC156, *Ventilation for Buildings*. The work of this group is divided into a number of subgroups. The situation is ongoing, and probably it is not helpful to give information here. The BSI is recommended as a good starting point for queries about the current state of play on the work of the TCs.

The progress of TC156 has been rather intermittent. Although progress reports are not routinely passed into the public domain, the consensus seems to be that whilst good progress has been made in the drafting and agreement of the so-called Bh horizontal standards, the progress of many of the groups dealing with more involved technical matters has been very slow.

For example, movement towards the concordance of agreement over ventilation requirements has been significantly hampered, amongst other things, by a protracted controversy about the validity of the olf system as a means of assessing ventilation requirements. When such difficulties are repeated to a greater or lesser extent in many of the other working groups, it is perhaps of little surprise that little progress is visible from the public domain.

Finally, the work of several other TCs may have some bearing on the matters of domestic ventilation. These include:

TC89 Thermal performance of buildings and building components.
TC126 Acoustic properties of building products and of buildings.
TC33 Doors, windows, shutters and building hardware.
TC166 Chimneys.
TC127 Fire safety in buildings.
TC48 Domestic gas-fired water heaters.
TC179 Gas-fired air heaters.
TC113 Heat pumps and air-conditioning units.
TC110 Heat exchangers.

References

1 F Engels. *The Condition of the Working Class in England.* Moscow: Foreign Language Publishing Houses, 1962.
2 Sir Parker Morris. *Homes for Today and Tomorrow,* 1961.
3 MJ Billington, MW Simons, JR Waters. *The Building Regulations Explained and Illustrated,* 12th edition. Blackwell Publishing, 2004. ISBN 0-632-05837-4.
4 Building Research Establishment. *Landfill Gas Remediation BR212.*
5 Building Research Establishment. *Radon Control in New Homes BR211.*
6 RR Walker, L Shao, M Woolliscroft. Natural ventilation via courtyards: theory and measurements. *14th AIVC Conference. Energy Impact of Ventilation and Air Infiltration* 1993, Copenhagen; pp. 235–250.
7 RK Stephen, LM Parkins, M Woolliscroft. Passive stack ventilation systems: design and installation. *BRE Information Paper 13/94* 1994, HMSO.
8 British Standards Institution. *BS5250: Code of Practice for the Control of Condensation in Buildings.* BSI, 1989.
9 British Standards Institution. *BS5250: Code of Practice for the Control of Condensation in Buildings.* BSI, 2002.
10 OFTEC Technical Note T1/112. *Oil Fired Appliances and Flues.*
11 British Standards Institution. *BS5440 Part 1 Spillage Testing.*
12 Building Research Establishment. BRE Digest 398, *Continuous Mechanical Extract Ventilation in Dwellings: Design Installation and Operation,* HMSO.
13 British Standards Institution. *BS5720: Code of Practice for Mechanical Ventilation and Air Conditioning in Buildings.* BSI, 1979.
14 British Standards Institution. *BS5925: Code of Practice for Ventilation Principles and Designing for Natural Ventilation.* BSI, 1991.
15 University College London Report for the *Department of the Environment, Transport and the Regions,* 1999.

16 E Perera, L Parkins. Airtightness of UK buildings: status and future possibilities. *Environmental Policy and Practice*, Vol. 2, No. 2. 1992; pp. 143–160.
17 MJ Billington, MW Simons, JR Waters. *The Building Regulations Explained and Illustrated*, 12th edition. Blackwell Publishing, 2004; pp. 14.11–14.22. ISBN 0-632-05837-4.
18 The Building, Northern Ireland Regulations, 1995–1997.
19 Technical Standards for Compliance with the Building Standards (Scotland) Regulations, the Scottish Office, 1995.
20 British Standards Organisation. *BS5588*, Part 9.
21 British Standards Organisation. *BS5925: Code of Practice for Ventilation Principles and Designing for Natural Ventilation*, 2nd edition, 1991.

8

Domestic Ventilation: Future Trends

8.1 Needs

The nature of the UK housing stock is very likely to change in order to reflect the demographical changes which are taking place; indeed, the signs are that changes which are already taking place. Trends towards single-person occupancy and increasing provision of sheltered accommodation for the elderly, coupled with ever-increasing demands on land are inevitably leading to the construction of smaller dwellings. Of these, an increasing proportion is likely to be flats in multiple-occupancy premises. At present, many suburban sites with large single dwellings are being redeveloped into relatively small apartment blocks. In addition, in cities such as Manchester, disused commercial buildings, such as warehouses and spinning mills, are finding a new lease of life as dwellings on a larger scale to those encountered in the suburbs. After many years of migration of citizens to the suburbs, this represents a major turnaround on public attitudes to city-centred life. Several estimates have been made which suggest that the population of central Manchester may grow by over 20,000 in the next 5 years.

This may be a timely occasion for a careful look at the efficacy of the possible ventilation strategies for flats and small dwelling units. Central mechanical systems are a logical first choice for blocks of flats. However, work is currently in progress to see if natural ventilation solutions can be successfully extended to such properties. It will be recalled from Chapter 6 that passive stack systems (PSV) have been installed in tower blocks in several European countries with some success. These made use of release of individual systems into a central discharge using the so-called shunt ducting. It will be necessary to find out whether these methods of ventilation can be modified for use within the UK. In particular, it will be interesting to see whether concerns about cross-contamination between flats can be allayed. Some of the pressure distributions around blocks of flats, particularly the taller ones, will be complex, and it may prove problematic to maintain extraction from all PSV simultaneously.

Overall, there are problems in the UK regarding supply of housing. In some areas, patterns of construction are already changing to try and meet demand, as has been described above. However, demand within the south-east in general and within London in particular, continues to grow, outstripping that within the rest of the country. The resultant rising property prices are putting even the simplest housing close to the workplace beyond the reach of workers. These people are compelled to commute long distances in order to have access to affordable housing. The problem has become so acute that at the time of writing this book it has become very difficult to fill public sector vacancies within the London area. In response to this problem, the government has signalled its intention to embark on the construction of what it calls "key-worker" accommodation. In order to make any impact on the problem, a large number of these properties must become available in a very short time. This brings the government up against another issue, namely that of the *availability of skilled labour*. At present, unemployment is low, and there is already considerable demand for labour within the construction sector. Indeed, given the large number of people who left the construction industry during the slump of the early 1990s, coupled with the almost total demise of the apprenticeship system, there is an acute shortage of skilled trade's people. Anyone who has recently tried to hire a reputable builder, plasterer or plumber to carry out some work on their house will testify to this. This means that the government's objective of the rapid construction of a large number of dwellings using more established methods is not possible. There has been a minor revival in the use of timber-framed construction, but it seems that the memory of the large-scale technical problems experienced in the dwellings built in the late 1970s and early 1980s is still strong. Some lenders are most reluctant to give mortgages on timber-framed properties. This is rather unfair in the case of more modern properties, since the technical issues are now far better understood and the quality of the timber-frame kits being used seems to be much higher than used to be the case.

For too many years, the public perception of construction has been of a dirty, inefficient industry employing low-quality staff. This is largely unfair, but image is of course very important in these days. Underlying this public perception is the undeniable and uncomfortable truth that the construction industry in the UK is nowhere near as efficient or productive as it could be, as it has an ingrained conservatism and resistance to the introduction of new methods and practices. As a result of considerable efforts by successive governments, the attention of the construction industry in general is finally being turned towards better productivity and quality of product, as promoted by the Latham and Egan reports. At the moment, there seems to be considerable commitment within the industry to pursuing the principles expounded within these documents. Whether this commitment would be a recession is a debatable point: it is to be hoped that the nation is not put in a position where it finds out. The modified stance of the construction industry to be expected, given the potential for higher profits arising from the use of more efficient business processes, coupled with a tight squeeze on margins. Part of the commitment involves a very careful rethink of the very way that buildings are constructed. In view of the shortages of the skills mentioned above, this is the most timely.

8.2 Technical

Much of the resultant process changes have been in the areas linked to the overall management of the construction process, for example, with respect to the management of the supply chain. By comparison, changes to the well-established sequence of on-site activities have perhaps not received as much attention. In Japan, some progress has been made in the use of robotics in construction: for example, in the laying of brickwork. The associated technical issues are still significant, and the widespread use of robotics would still seem to be a long way off. Significant changes are being seen in one key area. There is evidence for the introduction of factory methods into the construction process, not merely in the production of components (such as doors and windows), but also for much larger sub-assemblies. A significant part of the process is moving the indoors in the form of modular construction of rooms and segments of buildings. Many advantages associated with more established factory-based production activities may accrue. Repeatability should be easy to achieve. Delays in production due to adverse weather conditions should be massively reduced, since a very large proportion of the production process is now taking place indoors. The only vulnerable stages are raw materials or components delivery to factory, module delivery to site and a limited range of unavoidable on-site activities, such as ground works, services laying and connection, and the joining of individual modules into whole buildings. Surprisingly, many of the usual finishing requirements of buildings can be eliminated. It is possible for a contractor to purchase modular units that are not only decorated but also even furnished. Bathrooms and kitchen modules may be procured, which only need plumbing into site services once located in position.

Modular construction is already being used in the commercial sector. For example, most of the new restaurants built for the world's largest burger chain are supplied as fully fitted out modules. The time for completion of these construction projects may be as low as 6 weeks. In several industrial sectors such as pharmaceutical production, the use of modular facilities is becoming more common. In the residential sector, it is quite clear that the key-worker programme will be the main driver for the large-scale use of modular dwellings. Several companies are gearing up so as to be in a position to compete for a share of what should be a very lucrative market. However, at the time of writing there seems to be little evidence of a major acceleration in the erection of dwellings based on modules, although accommodation modules are finding great favour in the construction of small roadside motels.

The ventilation of modular units requires most careful consideration. Firstly, it was thought that it should be a simple matter to install trickle ventilators and supplementary ventilation in accordance with the requirements of Approved Document F; however, things are not that simple. Within modular units, there is a tendency towards the use of panel materials. This, coupled with the ease of assembly within the factory environment, leads to a large improvement of the airtightness of modular units in comparison to traditional building methods. These improvements go far beyond those achievable by the application of good practice in the traditional process. Air-leakage rates are being claimed which are more than two orders of magnitude smaller than those likely to be introduced as a result of future amendments to Building Regulations (more about this will be discussed later). This would have serious implications for the operation of PSV systems.

In recent years, there have been a number of developments in the technologies available for use within dwellings. The advances are probably not as significant in comparison to those in commercial construction (e.g. in passive solar architecture and natural ventilation for office buildings), but have nonetheless made a contribution. Equally, some (such as Trombe walls) have come and gone without much fuss. Perhaps their time might come again.

The future is of course a matter for speculation. Some ideas might seem too ludicrous for words; on the other hand, a close approximation to the hand-held communicator used by Captain Kirk and his crew in Star Trek is now on sale on the local high street in the form of the mobile telephone with video capability (admittedly, matter transportation is proving a tougher nut to crack). For big improvements to ventilation, indoor environment and energy conservation in dwellings, a look forward to some very new ideas is in order.

Some of the new ideas are already around, but do not seem to be very well publicised. For a showcase of new ideas, the author recommends a look at any publications emanating from the futuristic test house built by the University of Nottingham School of Architecture in collaboration with a consortium of companies. This work is ongoing, so more publications are likely to be pending. Other work is in progress on similar projects: for example, the Building Research Establishments (BRE) Eco Homes programme (which has a wider sustainability remit) and the "40% Homes study".

There are several technologies that look attractive. Photovoltaics are probably now capable of producing the levels of output necessary to power the size of fan, and essential to provide appropriate levels of ventilation.

For some of the newer high-efficiency fans, power consumptions of as little as 5 W are claimed. Most houses in the UK have pitched roofs, so there is great potential in theory for the placement of large areas of cells with no real change to the appearance of dwellings. The packaging of the photovoltaic cells is now much more robust than was the case with the early versions. Even better, the packaging forms may now be unobtrusive, for example, in the form of embedded cells within the tiles and slates themselves. This overcomes any issues of the appearance of the early photocell arrays, and also reduces the risk of *in situ* damage to cells. The tiles would be made of an epoxy or similar resin, in combination with fillers and colouring. The tile technology is well established: for example, Redlands have been making substitute slate tiles using ground slate waste for approximately two decades. The only unavoidable issue is that of the extra wiring behind the roof surface, which would exist for any photovoltaic installation. As the unit cost of photovoltaic cells decreases, the utilisation of the technology will undoubtedly increase, provided that the durability of the photocell carrying roof coverings have a comparable lifespan to traditional roof coverings, and also assuming that houses are built on an appropriate orientation for the maximisation of solar gain to the roof surfaces.

An important issue relating to energy efficiency is that of the basic philosophy of construction employed to dwellings. At one end of the scale, there are working examples of alternative designs that have been demonstrated to be very energy efficient. A good example would be that of the earth home, where much of the living space is underground. Such dwellings offer increased energy conservation coupled with more stable internal environmental conditions. Whether such dwellings would ever have mass appeal is questionable. Passive solar architecture is now well understood, and its principles are regularly employed in the construction of commercial buildings. A walk around a new housing development would raise doubts as to whether the same principles were being considered when houses were being designed. There is clearly much scope for the application of passive solar design techniques to the dwellings. Whether this will happen on the basis of developers being converted to the cause is another matter. Changes may be required for Building Regulations in order to make sure that the use of passive solar architecture in dwellings is maximised.

8.3 Legislation

Global warming has taken centre stage as an environmental issue. Despite reservations about the accuracy of predictions about climate and rising sea levels (if not in the reality of the phenomenon itself) in certain scientific circles, the prevailing view is that a gamble should not be taken. Certainly there are enough indications of changes in climate to raise alarm.

Central to current thinking on the control of global warming is the reduction of fossil fuel consumption. The UK Government is committed to implementing the Kyoto Treaty. Whilst renewable energy sources will contribute an increasing proportion of the total energy supply within the UK, reductions in carbon dioxide (CO_2) emissions of the order of magnitude required (60% by the year 2050 is now being suggested by the UK Government) can only be assured by large reductions in energy consumption. An appropriate reduction in energy consumption in buildings will, therefore, be needed. Apart from the merits of reducing the rate of global warming, better energy conservation will also have another benefit in that fossil fuel reserves will last longer. In the current uncertain political climate, this benefit in itself might be seen as a compelling reason for improving standards of energy conservation. What a shame that this particular nettle was more firmly grasped in the 1970s.

Friendly persuasion will only go so far. Without any doubt, most people would support in principle measures intended to protect the environment, but all the evidence points to them not being willing to significantly modify their behaviour on a voluntary basis. It is inevitable that the drive for energy conservation will be spurred on by legislation. The European Union (EU) is pushing on apace with an Energy Conservation Directive that will be binding on all member states. The Directive will be enshrined in national building codes. Therefore, there is little doubt that future revisions of Building Regulations will reflect the need to reduce energy consumption and hence reduce CO_2 emissions. It is logical to assume that close attention will be paid to air infiltration and ventilation. It has been mentioned in Chapter 7 that the current version of Part L of the Building Regulations for England and Wales (introduced in April 2002) sets an air-leakage standard of no more than $10\,m^3/h\,m^2$ of building fabric for commercial buildings, with a declared intention of reducing the standard to $5\,m^3/h\,m^2$ in the future. It is the understanding of the author that at the time of writing a review of Approved Document L is already under way, with the objective of quickly bringing forward a set of revisions that would bring Approved Document L in line with the requirements of the new EU Directive. It is inconceivable that such a review would not lead to the eventual (and possibly quite speedy) application of a prescriptive air-leakage standard to the dwellings. When this happens, there will be some critical issues to address with respect to ventilation strategies. As has been demonstrated in Chapter 7, an airtightness of $10\,m^3/h\,m^2$ within a typical dwelling would imply that the current Approved Document F requirement for trickle ventilator provision would no longer give sufficient open area to allow passive stack ducts sized in accordance with current good practice to draw air at a rate consistent with achieving control of condensation and mould growth; for that matter, the performance of extract fans would also be compromised, with the likelihood of short circuiting of air within the fan casing being greatly increased. Increase in air-inlet area would probably be needed in order to increase air flows in passive ducts to an acceptable level. Such an adjustment to open area sizes might give rise to concerns about increasing complaints from occupants about cold draughts within dwellings during the heating season. The effects of any proposed changes to open areas will have to be carefully studied prior to implementation.

It is hard to escape the conclusion that the imposition of a prescriptive airtightness standard for dwellings would push the construction industry towards the use of the centralised mechanical ventilation systems, providing both supply and extract air and almost certainly employing heat recovery from the exhaust air stream. As explained in Chapter 7, high levels of building airtightness leads to better coefficients of performance for such systems. However, in the opinion of the author, such a move would be regrettable for several reasons. It cannot be denied that the fan-manufacturing industry has made important technical improvements to its products which means that the power consumption of a fan of a size useful within a dwelling is likely to be very much lower than its equivalent of 10 years ago. This improved efficiency in turn improves the coefficients of performance of heat-recovery units in which the new fans are fitted.

Therefore, on first sight it might seem that the prospects for mechanical ventilation with heat recovery (MVHR) are much better. However, there are still serious concerns. Firstly, even though fans are now more energy efficient, they still consume electricity, and mechanical systems must be allowed to run continuously. If MVHR was to find its way into 15 million units of the current dwelling stock (about 75%), and two 5 W fans were to be used in each MVHR unit, then the associated daily additional power consumption of 150 MW would be incurred, corresponding to 197.1 TW per annum, in CO_2 emissions.

The use of mechanical systems has implications for the design, construction and maintenance of dwellings. The use of extract fans has few associated design issues beyond appropriate location and supply of electricity. The use of PSV means that space allowance must be made for ducts rising from ground floor level to the roof space prior to discharge; furthermore, the siting of extract points within kitchens, bathrooms and utility rooms relative to the location of the exhaust point must be carefully considered so as to avoid the use of inappropriate bends within duct runs. With balanced mechanical ventilation, the duct system needed is more extensive and complex, and will involve the use of risers and trunking. This will lead to a greater loss of space than with any of the other ventilation options. The design of the duct system should really be done at an early stage, or else difficulties may arise. Refurbishment projects may be problematic. Maintenance requirements are higher, indeed in some systems annual work will be needed. It is acknowledged that the other aspects of the new Approved Document L concerns building logbooks and a move towards management of technology (MOT)-style testing of commercial buildings. Whilst such measures should help to maintain the performance of commercial buildings, it seems questionable whether such measures could be introduced (and more importantly enforced) within the residential sector, especially in the light of the fact that currently it is considered to be neither practicable nor worthwhile to have a mandatory system of inspection for all gas appliances, which is unfortunately given the disturbingly high incidence of fatalities related to faulty-gas appliances.

Finally there is the issue of the objectives of the April 2002 version Approved Document L. After the first (and rather weak) attempt to discourage the use of mechanical ventilation and air conditioning in non-domestic buildings made in the 1995 version of the document, the intentions of the new document are quite clear. In the future, it will become more difficult to justify the use of mechanical systems in non-domestic buildings. Given this clear intention, it would be reasonable to assume that the government would in the fullness of time wish to pursue the same course of action for dwellings. If the air-leakage standards implemented and planned for the domestic buildings are applied to dwellings, then there is a danger that the consequence will be a swing towards the presumption of MVHR as the system of choice for new dwellings. Given the market positions being taken by certain manufacturers, it does not take much imagination to see a further consequence being a dramatic rise in the use of domestic air conditioning, as consumers would quite logically see air conditioning as the next step up the ladder. Surely these are the consequences that the government would not wish to arise as a reaction to its well-intentioned environmental policies. Perhaps a lower standard of airtightness for dwellings might be appropriate, sacrificing some controllability of ventilation and increasing air infiltration, but permitting non-electricity-consuming ventilation measures, to continue to perform effectively.

Other regulations affecting the ventilation of dwellings are in the process of being changed. Approved Document F is not at the time of writing thought to be due for imminent revision. However, there are currently some documents in circulation (but as per the author's knowledge it is not publicly available) emanating from the BRE, which provide some possible pointers to Government thinking on issues of domestic ventilation legislation in the future. It would appear that one possibility is that the historic focus of Approved Document F will be broadened from the control of condensation to a wider remit for the maintenance of satisfactory air quality. Furthermore, it seems that one particular area of concern is the control of nitrous oxides, primarily emitted as a by-product of the combustion of natural gas. The documents in question would have dramatic implications for kitchen ventilation in that natural or PSV strategies would struggle to provide the level of ventilation seemingly required. However, much has been made of the very limited body of experimental evidence being cited, coupled with dissent over the rather low concentrations of nitrogen oxides that are being put forward as a potential health hazard.

Important changes are being proposed to Approved Document C that will impinge on Approved Document F. A consultation document was issued in late 2002. Some changes to measure for radon control are proposed: for example, the introduction of sub-floor ventilation for suspended concrete floors. Of far more significance is the inclusion of clauses relating to the prevention of interstitial and roof-space condensation within the proposed revisions. This would imply the removal of Part F2 from Approved Document F. It would appear that the intention of the Government is to transform Approved Document F into something more related to other indoor air-quality issues. It will be interesting to see what future is proposed for Part F when the next revisions are published. As a matter of interest, the proposals for the control of condensation within pitched roofs seek to sanction the use of unventilated roof spaces, instead relies on vapour permeable felt. This issue is discussed in further detail in Chapter 6, and will not be repeated here. Suffice it to say that there are significant misgivings in various quarters about this specific proposal for a variety of reason, not the least of which is the relative lack of medium-to long-term experimental data to back it up. A major producer of breathable membranes has recently been ordered to modify the claims made in its advertisements regarding the applicability of such membranes in domestic roof spaces. The difficulties have in the most part arisen in the interpretation of the term "unventilated". The membrane manufacturers seem to interpret the term in its most literal sense, but the majority view is that ventilation of the batten space beneath the tile covering is still essential in order to control damaging condensation. This is entirely logical, given that if the membrane is functioning as intended, then water vapour will be free to permeate into this cavity.

There is one final issue to be considered with respect to Building Regulations. This is the introduction of a hitherto uncommon concept, namely that of performance-based regulations. The concept of performance-based regulations involves the setting of specific performance standards for specific aspects of building performance, but from the viewpoint of the user requirement. This is a big move away from the usual way of regulation construction, which relies on the internationally widespread practice of basing building codes and regulations on the prescriptive ("deemed to comply") specification of requirements for building materials, components and systems. Prior and Szegeti[1] discuss the principles behind the philosophy of the performance approach to construction. They note that the code-based approach may or may not "provide a good value for money construction which is fit for its purpose". Furthermore, they point out that "deemed-to-satisfy" codes tend to concentrate on a material solution to a perceived issue that may not have been properly understood in the first place, or for that matter not even clearly defined. The code-based approach is presented as an obstacle to innovation, cost optimisation and even trade. Prior and Szegeti present the application of performance-based regulations as a logical means of achieving a range of benefits in the construction process: for example, the promotion of shared interest between clients and construction businesses.

Prior and Szegeti summarise progress on the performance-based approach, mentioning the work of the USA General Services Administration of the last 50 years, the work of the Canadian Centre for Facilities that has taken place for 30 years and the foundation by the CIB of the PeBBu network in 2001. It is pointed out that several European governments are actively pursuing research in the area. Prior and Szegeti claim that the UK Building Regulations (presumably for all national regions) are "predicated on a performance basis, and are supported by a series of Approved Documents in which helpful guidance is given in the form of deemed to satisfy examples".

In areas such as Part L (Conservation of Fuel and Power), the validity of this statement could reasonably be said to be clear. However, the author questions whether that could be said to be the case for Part F, where although the actual regulations F1 and F2 are arguably performance based, demonstration of compliance is centred around prescriptive requirements for open areas of air inlet, fan-extraction rates, air-change rates for mechanical systems and the like. Given the emphasis on the control of condensation, the current Approved Document F might be viewed as falling into trap of tending to (in the words of Prior and Szegeti) "concentrate upon a material solution to a perceived issue that may not have been properly understood in the first place ...". It would surely be expected that performance-based regulations would lay down specific targets for such parameters as indoor relative humidity and other air pollutants. It is most interesting to note that at a recent seminar, the BRE gave a very interesting presentation about their current work on ventilation. A key theme of this presentation was the intention to move towards a performance-based legislative framework for ventilation within buildings including dwellings. Therefore, it would be logical to assume that the next version of Part F would take such a shape.

8.4 Conclusions

Housing within the UK continues to evolve, as does the demands of the population. In terms of future trends, it is clear that global warming and sustainability will be the principal drivers, closely followed by demographical pressures of household sizes and land availability. Changing housing will inevitably be accompanied by changing ventilation strategies.

The author would like to be able to make the confident prediction that the process of change will lead to a massive reduction in the number of dwellings affected by the most fundamental ventilation-related issue, namely condensation. Despite well over 30 years of prescriptive legislation intended to lead to a decreased risk of condensation and associated mould growth in dwellings, the findings of the most recent English House Condition Survey make depressing reading. In Chapter 1, reference was made to the book edited by Croome and Sherratt. The Building Regulations may be different, the *U*-values are higher; but it is quite clear that the researchers of the day had a good understanding not only of the causes of condensation in dwellings, but also of several of the possible causes. One might question how many lessons have in fact been learned in the interim.

It is to be hoped that future legislative changes will not only lead to the urgently needed improvement, but will also address the serious problems within the existing stock. One wonders what Edwin Chadwick and the other Victorian campaigners would make of our world: a world of contradictions, a world in which overall living standards are high and yet a basic, entirely avoidable health-related issue is allowed to persist despite the best efforts of scientists, engineers, civil servants and politicians. Too many cooks? ... or merely the wrong recipe book?

References

1 J Prior, S Szegeti. Why the fuss about performance based building regulations? Paper presented to *CIB Working Commission on Performance Based Building Regulations*, Budapest, March 2003.

Index

Page numbers in *italics* refer to figures and tables.